Puzzled by Math!
Using Puzzles to Teach Math Skills

by Trish Caldwell Landsittel

Prufrock Press, Inc.
P.O. Box 8813
Waco, Texas 76714-8813
(800) 998-2208
Fax (800) 240-0333
http://www.prufrock.com

Table of Contents

Acknowledgements iv

Introduction 1

Puzzles

Two Addends 3

Three Addends 4

Addition: Mental Math 5

Estimate the Sum 6

Addition Practice 7

Subtracting by 1 Digit 8

Subtracting by 2 Digits 9

Subtracting by 3 Digits 10

Estimate the Difference 11

Subtraction Practice 12

Multiplication by 1s, 2s, 3s, & 4s .13

Multiplication by 5s, 6s, 7s, & 8s .14

Multiplication by 9s, 10s, 11s, & 12s 15

Multiplication by 1 Digit 16

Multiplication by 2 Digits 17

Division by 1s, 2s, 3s, & 4s 18

Division by 5s, 6s, 7s, & 8s 19

Division by 9s, 10s, 11s, & 12s 20

Division by 1 Digit (With Remainder) 21

Division by 2 Digits (With Remainder) 22

Adding Fractions (Same Denominator) 23

Subtracting Fractions (Same Denominator) 24

Adding Fractions (Different Denominators) 25

Subtracting Fractions (Different Denominator) 26

Reducing Fractions to Simplest Form 27

Adding Decimals 28

Subtracting Decimals 29

Multiplying Decimals 30

Fraction and Decimal Conversion .31

Comparing Decimals (Greater Than, Less Than,
and Equal To) 32

Basic Algebra 33

Algebra: Addition and Subtraction 34

Squares 35

Cubes 36

Square Roots 37

Answer Key

Solution: Two Addends 38

Solution: Three Addends 38

Solution: Addition: Mental Math .39

Solution: Estimate the Sum 39

Solution: Addition Practice 40

Solution: Subtracting by 1 digit 40

Solution: Subtracting by 2 digits .41

Solution: Subtracting by 3 digits .41

Solution: Estimate the Difference .42

Solution: Subtraction Practice 42

Solution: Multiplication by 1s, 2s, 3s, & 4s 43

Solution: Multiplication by 5s, 6s, 7s, & 8s 43

Solution: Multiplication by 9s, 10s, 11s, & 12s 44

Solution: Multiplication by 1 Digit 44

Solution: Multiplication by 2 Digits 45

Solution: Division by 1s, 2s, 3s, & 4s 45

Solution: Division by 5s, 6s, 7s, & 8s 46

Solution: Division by 9s, 10s, 11s, & 12s 46

Solution: Division by 1 Digit (With Remainder) 47

Solution: Division by 2 Digits (With Remainder) 47

Solution: Adding Fractions (Same Denominator) 48

Solution: Subracting Fractions (Same Denominator) 48

Solution: Adding Fractions (Different Denominator) 49

Solution: Subtracting Fractions (Different Denominator) . . .49

Solution: Reducing Fractions to Simplest Form 50

Solution: Adding Decimals 50

Solution: Subtracting Decimals 51

Solution: Multiplying Decimals 51

Solution: Fraction and Decimal Conversion 52

Solution: Comparing Decimals (Greater Than, Less Than,
and Equal To) 52

Solution: Basic Algebra 53

Solution: Algebra: Addition and Subtraction 53

Solution: Squares 54

Solution: Cubes 54

Solution: Square Roots 55

Appendix—Math Resources 56

Blank Puzzle: 5 x 5 57

Blank Puzzle: 5 x 6 58

Blank Puzzle: 6 x 7 59

Acknowledgements

Two individuals were instrumental in seeing my dream come to fruition: Jennifer Robins, editor, and Sharon Meyer, 3rd grade teacher with Barnesville Elementary. Through an endless array of e-mails, Jenny tirelessly shared her expertise, depicted perfect patience, provided feedback through lightening-speed responses, and always displayed the most diplomatic approach with her suggestions. Sharon, whose tech-savvy skills created the grids and the means for me to manipulate the data, embraced the opportunity to troubleshoot, research the best solution to a never-ending array of dilemmas, and create a countless number of grids before "we" settled on the ones presented in this publication. You're right Sharon. It was a true "partnership." I will forever be grateful to the two of you.

To my ultimate teacher: my son, Ben Landsittel. The education I gained in being your mom has exceeded any one educational endeavor. You are my greatest blessing. To my mother, Arlene Caldwell, who defines the word perseverance—seeing you rise above more obstacles than any one individual should ever have to endure, you constantly demonstrate that emotional strength can supersede anything imaginable. To my "favorite brother," James Caldwell, who has never allowed others to limit his potential—you've paved more than one road for me to follow. To my sister, Donna Potts: you were the first to teach me that siblings can be best of friends. It's through the unique contributions of these "fabulous four" that I'm able to achieve my dream. Thank you for your unconditional, never-ending supply of love.

Thank you to the following individuals who made unique contributions—sometimes unknowingly: Lisa Stupak, Janice Milliken, Mike Marsh, Anne Wright, Gail Swingle, Pam Griffith, Rena Allen, Paula Ball, Karen Brown, Jennifer Cullen, Johanna Morris, Dorothy Templeton, Misty Rose, Tessa Liniger, Sharon Cain, Kirstin Repco, Tina Lees, Cheryl Potts, Reta Gundlach, Martha Ackerman, Trudy Mahoney, Karen Collins-Jarvis, Barnesville Elementary School Faculty and Staff, and Dr. Gupta.

And last, but certainly never least, thank you to the countless number of gifted students I've had the privilege of serving over the years (Switzerland of Ohio Riverfront TAG, Barnesville Elementary;.and Southwest Licking). Your honest and uninhibited feedback helped develop this product.

Introduction

Description of the Book

Puzzled by Math! is a collection of mathematical equations, knowledge, and skills placed in puzzle form. The puzzles provide advanced learners the enrichment they need through a self-paced, independent learning mode. Once basic instructions and pertinent materials are provided, students possess the ability to self-check and progress at their own pace to acquire increasingly more difficult math skills.

Directions

Students will need a puzzle and an appropriate sized, blank puzzle grid. Separate the puzzle pieces by cutting along the heavy, dotted, black lines. The name of each puzzle denotes the "matching" objective. Match the edge of one piece (the equation) with the edge of another (the solution). Continue in this manner until all pieces are matched forming a complete puzzle. Once all the pieces are matched correctly, glue them onto the proper grid location.

Please note: the student should not glue the pieces into their permanent location on the blank grid until after all the pieces have been manipulated. Too many blocks hold a first-appearance of being correct, only to discover later that such is not the case.

Summary of Content and Skills Found in Book

The content addresses many of the essential skills and knowledge contained in the National Council of Teachers of Mathematics Standards. Your students will:
- build new mathematical knowledge through problem solving;

- select and use various types of reasoning and methods of proof;
- understand numbers and various means of representing numbers;
- understand patterns, relations, and functions;
- analyze mathematical situations and structures utilizing algebraic symbols;
- evaluate inferences and predictions; and
- understand how mathematical ideas interconnect and build on one another to produce a coherent whole.

Although the curriculum is designed for grades 3–7, the mere presentation of the subject matter allows for students of lower grades as well as higher grades to utilize the materials contained in this text. This is due to several factors:

1. degree of difficulty can be adjusted by implementing mental-math or allowing a more mechanical approach;
2. with some blocks sharing a common "appearance" with others, the student is required to assess all four sides to determine proper placement; and
3. some concepts allow for a progression through mere size of the puzzle itself. This advancement brings about an increasingly more difficult challenge with more puzzle pieces to assess and manipulate.

A multitude of skills are called upon to solve each puzzle:
- with equations presented in a linear fashion, a student must hone his or her skills in transferring data into a more usable form (vertical pattern)—especially those students using mental math to solve the problems;

- in the initial stage, sorting, grouping, and categorizing are necessary skills the student must develop to help manipulate multiple pieces;
- organization and memory skills play an intricate part (please note: some students will readily observe and differentiate between border pieces and center pieces, helping with the ultimate solution while other students will not—it is the advice of the author not to verbalize this to the students—students need to acquire problem-solving skills through self-exploration); and
- because mathematical solutions can be reflected in a variety of forms, prediction plays a key role. A student will come to see and consequently predict the various possibilities for each equation, while remaining open to the appearance of the needed puzzle piece.

Because advanced learners acquire new knowledge and skills quickly and without repetition, very little duplication exists in the content of these puzzles.

Helpful Tips

Encourage independent learning. It's one of the greatest gifts we can instill in our youth. One approach is through learning centers. The materials presented in this book can be turned into a learning center quite easily. Duplicate numerous copies of each puzzle and the blank puzzle grids. Place each category in a labeled file folder within a box, container, or large expandable. Copy the answer keys on colored cardstock, laminate, and place on a ring in numerical order. Hang in close proximity of the math center. Laminated keys will endure repeated use and the ring will eliminate the possibility of lost solutions.

Establish clear expectations for the learning center (i.e., you may use the center when your work is complete; the answer keys are provided for you to self-check your work or gain direction if unclear about a possible solution, etc.).

As an extension, have students create their own math puzzles. Students can create an answer key as well as a student sheet comprised of scrambled blocks. Encourage students to be creative when naming their puzzles. After developing their own, have the students exchange puzzles with one another.

To reinforce new mathematical concepts being taught in the classroom, have students work puzzles in groups or as individuals. Make numerous copies of each puzzle on colored cardstock (a different color for each puzzle), laminate, cut apart, and place each one into its own labeled baggie. Rather than the usual paper and pencil method, students can practice their newly acquired skills through manipulation and puzzle creation. Let them practice repeatedly over a number of days competing for their own best time.

For additional information or suggestions, see the Appendix or feel free to contact the author at shamrockgirl_oh@yahoo.com.

Two Addends

245 + 168   203 + 40   31   20 + 10   16 + 7

32 + 13   19 + 9   12 + 6   5 + 16   95   15 + 10   85

82   173   278   48

200 + 15   26   41 + 7   250 + 18

55   23   41 + 54   100 + 21   12 + 11   18

216   23   92

25 + 11   47 + 35   127 + 46   74 + 12

25   44 + 11   12 + 11   21   23   18   14 + 7   72 + 13   12 + 15

243   224   8 + 12

202 + 14   289   400 + 86   14 + 7

32 + 45   8 + 15   9 + 12   15 + 9   24   8 + 10   77   21

17 + 9   268   36   285 + 4   30

64 + 28   250 + 28   20   214 + 10

21   27   28   23   21 + 10   121   16 + 8   21   45

413   486   86   21   215

Three Addends

11 + 47 + 23
100 + 24 + 89
87 + 17 + 245
188

75 + 68 + 33
47 + 34 + 98
1,668
96

43 + 17 + 41
727
141

25 + 35 + 10
74
176

25 + 35 + 3
777 + 888 + 3
81

736 + 87 + 9
12 + 23 + 56
33 + 32 + 10
123
160

13 + 13 + 13
101 + 45 + 123
56 + 77 + 10
126
40

2 + 50 + 60
143
75
87

79 + 34 + 13
199
396
70

14 + 25 + 121
16 + 13 + 78
349
134

111 + 22 + 8
177
101

11 + 2 + 27

36 + 24 + 1
53
159

120
5 + 56 + 13
152

25 + 35 + 10
98 + 160 + 34
25 + 20 + 8
119
20 + 40 + 60

52
17 + 8 + 34

76 + 78 + 34
832

33 + 21 + 22
188
112

61 + 13 + 78
59
213

77 + 90 + 21
179
76
39

89 + 21 + 13

8 + 9 + 16
259
10 + 8 + 34

23 + 80 + 56
333 + 33 + 30
157

50 + 25 + 12
45 + 87 + 45

190
56 + 78 + 125
61

61 + 24 + 11
77 + 34 + 17
91
269
103

83 + 34 + 17
2
28 + 78 + 84

25 + 57 + 75
292
70

62 + 25 + 16
676 + 34 + 17
107
74

42 + 21 + 11
99 + 56 + 44
$4 + $12 + $11
25 + 16 + 78
33

Addition: Mental Math

70		30		100
250 + 28	90 · 500 + 200 · 400 + 86	278 · 130	1,100 · 300 + 500	900 + 400
100 + 150	40 + 30	975	60 + 10	700 + 50
	70	370	80 + 20	250
700	80 + 50 · 1,700 · 110 + 90	268	1,000 + 700 · 243	
980	50 + 50	350		350 + 20
	35	35	90	450 + 50
700 + 400	300 · 1,300	486	619 · 214 + 10	300 + 150
25 + 10	50 + 20	500		45 + 10
80 + 006		500	900 + 75	100 + 250
200	250 + 28 · 40 + 50	200 + 15 · 450	203 + 40	800
25 + 10	100	40 + 50	400 + 86	400 + 100
70	50	55	486	750
278	150 + 150 · 407 + 12 · 23 + 23 · 46	70 + 50 · 120	250 + 18 · 215	224
15 + 15				20 + 30

Estimate the Sum

50

46 + 46

23 + 72

70

12 + 11

80

3 + 8

90

18 + 22

90

63 + 34

100

16 + 8

26 + 38

70

43 + 29

60

100

88 + 12

2 + 7

62 + 11

52 + 25

100

40

8 + 3

60

1 + 11

100

40

26 + 14

7 + 11

32 + 21

60

100

44 + 32

75 + 15

10

40

54 + 42

100

36 + 24

20

10

67 + 12

40

20

70

47 + 8

30

11 + 12

100

31 + 72

90

70

19 + 19

40

16 + 9

34 + 42

22 + 72

27 + 19

38 + 29

20

50

91 + 12

70

90

84 + 24

80

36 + 24

9 + 6

70

17 + 17

30

10

90

52 + 52

81 + 11

26 + 9

Addition Practice

898 · 108 + 12 · 766	959 · 150 + 50 · 279 + 63 · 289 + 46	891 · 203 + 45 · 121 + 680 · 116 + 133	345 · 178 · 400 + 332	345 + 345 · 560 + 56 · 327 + 372 · 36 + 12	671 · 699 · 763 · 845 + 23
432 + 275 · 231 + 141 · 342 · 164 + 348	64 + 75 · 103 + 75 · 110 + 346	48 · 96 + 12 · 120	676 + 46 · 772 · 204 · 732 + 159	663 · 745 + 18 · 112 + 132	300 + 500 · 342 · 93 + 14
415 · 801 · 508 · 467	352 · 830 · 777 + 111	512 · 11 + 111 · 217 · 259 + 217	677 · 820 · 248 · 438 + 219	101 + 103 · 139 · 707	107 · 250 + 28 · 137 + 248
521 + 234 · 182 + 35 · 439 + 333 · 428 + 249	323 + 144 · 284 + 394 · 525 + 125 · 295 + 376	671 · 650 · 436 + 227	888 · 980	227 · 356 · 333 + 111	385 · 422 + 86 · 295 + 376
144 · 84 + 12 · 801	132 · 946 + 34 · 96	369 + 146 · 122 · 117 + 110	249 · 109 + 91 · 878 · 690	111 + 12 · 800	244 · 955 + 42
657 · 245 + 168 · 200 · 255 + 167	785 · 729 + 101 · 308 · 12 + 120	422 · 204 + 104 · 616 · 24 + 120	204 · 108	456 · 279 + 63 · 866 · 755	242 + 103 · 123 · 726 + 233
732 · 913 + 85 · 200 · 722	127 + 428 · 811 + 110 · 413 · 327 + 458	85 + 23 · 372 · 515	476 · 143 + 213 · 555	444 · 470 + 350 · 921 · 231 + 121	335 · 151 + 53 · 278 · 358 + 57

Subtracting by 1 Digit

729

2,080 - 3

862

220

306

289 - 3 | 1,050 - 5 | 77 - 5 | 44 | 91 | 1,045 | 25 | 72 | 32

991 | 612 - 9 | 5,379 - 4 | 462 - 2 | 320 - 8

166 | 5,375 | 172 | 361

48 - 4 | 64 - 2 | 996 - 3 | 75 - 3 | 447 - 5 | 288 - 7 | 27 - 3

364 - 3 | 795 | 225 - 5 | 180 - 8

981 | 871 | 608 | 631

286 | 749 - 4 | 51 | 363 - 5 | 993 | 559 | 281 | 197 - 5 | 745

673 - 7 | 732 - 3 | 813 - 4

999 | 997 - 6 | 460

36 - 4 | 32 | 211 | 217 - 6 | 72 | 568 - 9 | 24 | 62

628 - 8 | 635 - 4 | 878 - 7 | 385 - 4 | 315 - 9

620 | 312 | 798 - 3 | 603 | 381

192 | 28 - 3 | 95 - 4 | 36 - 4 | 54 - 3 | 442 | 358

2,077 | 866 - 4 | 987 - 6

Subtracting by 2 Digits

441	2,080 - 40	832	801	100
217 - 13	757 -35	795 - 24	973 441	726
722	714	702		
870 - 69	601	5,149 - 34	210	
5,115		610	673 - 62	
748 - 34	220	740	554	271
		145	360 - 24	
364 - 12	315 - 10		628 - 18	802
975		612 - 11	302	305
288 - 17	504 - 63	447 - 25	757 - 31	536 - 18
276	275 - 55	265	771	
611	180 - 80		2,040	320 - 18
352	985	798 - 31		371
422	336	518 204	568 - 14	343
	635 - 44	866 - 34	385 - 14	987 - 12
701	591	225 - 15		813 - 11
289 - 13	197 - 52	749 - 47 / 1,050 - 25	764 - 24 / 288 - 23	368 - 25 / 996 - 23
1,025				
997 - 12		462 - 21	767	732 - 31

Subtracting by 3 Digits

803 - 411
368 - 250 996 - 223
732 - 231

444
795 - 249
5,149 - 768

462 - 253
217 - 137
870 - 369

$278 - $155
$757 - $109
118
676 - 323

1,594
$648
192
612 - 307

4,381
748 - 556
364 - 125

280
$623
191
305 - 180

239
288 - 172
7,200

504 - 463
555 - 275

305
447 - 256

501
749 - 467
41
153
997 - 182

501
825
1,050 - 225
209

45
56

635 - 244
197 - 152
282

426
$123

320 - 202
288 - 232
545
568 - 142
2,000 - 406

116
354
392

798 - 311
866 - 422

448
45
487

$764 - $141

125
757 - 123
536 - 182
118

815
360 - 242
80
391

144
118
628 - 180

8,000 - 800
773
634
210 - 165

353
289 - 136
498 - 354

Estimate the Difference

30	10	20	74 - 25	70
37 - 12 / 20 / 91 - 22	20	35 - 8	49 - 4 / 70	10
81 - 6	10	75 - 9		53 - 24
08	20	50	40	92 - 55
20	22 - 3	10	80 / 16 - 8	43 - 29
	78 - 3		26 - 14	36 - 24
20			30	77 - 6
70 / 30 / 38 - 29		40	44 - 32 / 88 - 12	22 - 18
40	20	72 - 31		24 - 16
48 - 41	70			
33 - 18 / 101 - 34 / 80 - 13	47 - 34 / 50	10 / 64 - 11	72 - 22 / 54 - 13	50
65 - 15		34 - 17	63 - 34	70
0	28 - 14	70	48 - 41	30
30 / 10 / 52 - 25	67 - 12	10	60 / 0	70
10	79 - 8	56 - 38	41 - 14	30

Subtraction Practice

179 394 - 204 525 - 125 876 - 347	933 18 190 345 - 45	304 400 436 - 222	90 84 - 12 12 - 96	259 - 217 213 - 79 20 498 - 34	123 100 504 120 - 30
630 439 - 333 115 732 - 159	911 250 - 28 248 - 36	213 913	89 913 - 85	70 946 - 34 100 72 676 - 46	231 231 - 121 279 - 63 348 - 160
369 - 146 111-11 100 117 - 110	188 127 42	408 - 106 597 - 324 432 - 201	777 - 111 916	243 168 - 53 222 301	110 628 666 999
7 134 333 - 111	300 560 - 56 171 36 - 12	273 122 195 - 45	573 203 - 45 956 - 23	358 - 57 680 - 120 560 429 - 16 323 - 144	150 216 828 287
118 - 12 200	464 701 794 - 525	219 77 245 - 168 255 - 132	109 - 16 808 - 80	222 302 231 - 121	728 811 - 110 110 223
521 - 234 182 - 55 428 - 249	269 106 729 - 101 120 - 50	822 204 - 104 96	529 242 - 103 955 - 42 726 - 233	372 - 201 106 732 845 - 23	214 745 - 13 999 - 786
500 - 300 216 922 - 11	493 193 - 71 139 400 - 332	150 - 50 289 - 46	24 279 - 63 84	212 108 - 12 413 376 - 72	179 470 - 450 158 438 - 219

Multiplication by 1s, 2s, 3s, & 4s

3 × 10   4 × 8   0 × 20   40   22

1 × 3   10   1 × 12   9   4 × 3   2 × 7 / 1 / 3   1 × 7?   2 × 6

2 × 10   4 × 11   4 × 0

20   20   6   3 × 8   2

16   4   12   1 × 2   1 × 7   1 × 9 / 1   18

2 × 11   0   16   30   3 × 5

3 × 9   4 × 4   4 × 9

1 × 8   7   2 × 5 / 2 × 9   3 / 6   12

3 × 4   3 × 6   8   3 × 11   4 × 10

4 × 12   15

10   4   1 × 9   24   1 × 11   2   1 × 4   11   1 × 5   8

4 × 1   3 × 3   27   32

21   33   9   18   4 × 2

2 × 1 / 1 × 10   1 × 10   2 × 2   24   2 × 4   2 × 8

3 × 12   3 × 2   4 × 5   3 × 7   4 × 7

28   44   36   0   4

2 × 3   14 / 3   6   5   4 × 6   12   2 × 12   8

48   36   24

Multiplication by 5s, 6s, 7s, & 8s

96	63	20	0	7 × 12
6 × 2 6 × 8 / 6 × 9 6 × 1		24 5 × 8 / 8 × 1	5 × 7 18 72	9 6 × 9
32 8 × 8	5 × 8 8 × 8 / 8 × 5	72 8 × 1	21 88	56 6 × 9
40 48	42 56	6 × 10	36 30	50
7 × 6 7 × 11	54	24	40 35	7 × 7 70
5 × 10 84	25 6 × 6 8 × 10	55 7 × 5	5 × 9 7 × 2	5 × 5 5 × 1 7 × 9
30 14	10 77	28 6 × 3 7 × 3	49 6 × 12 8 × 3	42 6 × 4 5 × 12 7 × 8
5 × 3 8 × 6	45 66 8 × 11 8 × 12	60 5 7 × 1	7 × 4 5 × 4	48 5 × 6 5 × 11 15 64 8 × 9
8 × 10	12 5 × 2 7	60 8 8 × 0	6 × 5 6 × 9 80 6 × 7	35 8 × 7 6 × 11 80

Multiplication by 9s, 10s, 11s, & 12s

Multiplication by 1 Digit

Row 1:
- 23 × 8 | 28 × 8 | 486 | 41 × 4
- 432 | 111 × 0 | 30 × 4
- 288 | 32 × 7 | 34 × 4
- 243 | 1,256 | 306
- 164 | 88 | 110

Row 2:
- 70 × 7 | 102 | 54 × 9 | 40 × 8
- 350 | 520
- 320 | 192 | 44 × 2 | 60 × 9
- 130 × 4 | 3 × 34 | 72 × 4
- 270 | 224 | 68

Row 3:
- 24 × 2 | 62 × 5 | 102 × 3 | 7 × 25 | 54 × 9
- 7 × 101
- 540 | 34 × 6 | 24 × 5 | 54 × 8
- 486 | 399
- 4,992 | 490

Row 4:
- 6 × 10 | 24 × 3 | 306 | 81 × 3
- 272 | 310 | 72 | 30 × 4
- 120 | 175 | 120 | 25 × 8
- 120 | 120 | 880 | 48
- 136 | 224

Row 5:
- 34 × 2 | 6 × 32 | 624 × 8 | 220 × 4 | 68 × 4
- 96 | 60 | 30 × 9
- 707 | 50 × 7
- 22 × 5 | 133 × 3
- 200 | 628 × 2 | 0 | 4 × 24 | 184

Multiplication by 2 Digits

1,681	2,880	156	68 × 48	34 × 43
1,224	44 × 15 / 111 × 50	660	748	220 × 17 / 4,650	364
22 × 15		72 × 40	45 × 12	24 × 28
105 × 25	3,780	81 × 23 / 105 × 10 / 8,792	133 × 40 / 288 / 330	5,550
	3,690 / 672			
480		1,462	1,863	
25 × 15	408	5,940	1,600	
628 × 14	42 × 32 / 800,1	3,740 / 34 × 22	624 × 34 / 2,916	
23 × 18	3,264	54 × 70	76 × 74	13 × 12
21,216	5,624	4,136	40 × 12	324
	1,050	1,224	42 × 24 / 1,344	960	5,320
30 × 93	47 × 88	60 × 99	50 × 32	
414	540	2,790	672	
28 × 13	54 × 54 / 123 × 30 / 54	24 × 40 / 36 × 34	24 × 12 / 32 × 21	62 × 75 / 2,625	102 × 12
41 × 41		34 × 12	375	27 × 12

© 2005 Trish Caldwell Landsittel

Division by 1s, 2s, 3s, & 4s

Division by 5s, 6s, 7s, & 8s

8 · 12 · 14 ÷ 7 · 40 ÷ 8 · 120 ÷ 6 · 13 · 48 ÷ 8 · 45 ÷ 5 · 56 ÷ 7 · 4 · 16

9 · 7 · 77 ÷ 7 · 20 ÷ 5 · 75 ÷ 5 · 15 · 20 · 10 ÷ 5 · 4

18 ÷ 6 · 72 ÷ 8 · 21 ÷ 7 · 24 ÷ 8 · 84 ÷ 7

30 ÷ 5 · 85 ÷ 5 · 25 ÷ 5 · 65 ÷ 5 · 54 ÷ 6

14 · 2 · 55 ÷ 5

56 ÷ 8 · 2 · 20 · 3 · 60 ÷ 5 · 3 · 10 · 3

72 ÷ 6 · 10 · 64 ÷ 8

35 ÷ 5 · 8 · 9 · 3 · 5

10 · 2 · 8 · 63 ÷ 7 · 9 · 70 ÷ 7 · 50 ÷ 5 · 16 ÷ 8

66 ÷ 9 · 48 ÷ 6 · 18 · 9 · 80 ÷ 5

12 · 6 · 90 ÷ 5 · 11 · 60 ÷ 9

42 ÷ 6 · 5 · 42 ÷ 7 · 4 · 24 ÷ 6 · 28 ÷ 7 · 5 · 6

4 · 15 ÷ 5 · 70 ÷ 5 · 5 · 11

8 · 7 · 19 · 17

100 ÷ 5 · 9 · 6 · 7 · 35 ÷ 7 · 30 ÷ 5 · 36 ÷ 6 · 7 · 32 ÷ 8 · 12

49 ÷ 7 · 11 · 40 ÷ 5 · 95 ÷ 5

3   10 ÷ 10   20   44 ÷ 11   11   3

12 ÷ 12   110 ÷ 11   121 ÷ 11   50 ÷ 10   22   120 ÷ 12   100 ÷ 10   132 ÷ 11   4   1

121 ÷ 11   4   100 ÷ 10   18 ÷ 9

120 ÷ 10   11   48 ÷ 12   24 ÷ 12   12   11

6   5   6   10   7   8   11   9   33 ÷ 11

110 ÷ 10   4   220 ÷ 11   2

84 ÷ 12   3   36 ÷ 9   45 ÷ 9   120 ÷ 10   110 ÷ 10   96 ÷ 12   3

6   8   5   10 ÷ 10   20 ÷ 10   2   6   108 ÷ 12

132 ÷ 12   12   9 ÷ 9   33 ÷ 11   60 ÷ 10

10   9   66 ÷ 11   90 ÷ 10   40 ÷ 10   77 ÷ 11   81 ÷ 9   4   132 ÷ 12   11

9   6   66 ÷ 11

11 ÷ 11   6   1   108 ÷ 12   108 ÷ 9   2

22 ÷ 11   1   22   1   2   1

264 ÷ 12   84 ÷ 12   48 ÷ 12   88 ÷ 11   10   18 ÷ 9   12   9   1,000

100 ÷ 10   66 ÷ 9   144 ÷ 12   1   264 ÷ 12

60 ÷ 10   7   10   12 ÷ 12

10   60 ÷ 12   12   9   3   11   8   7   11

20 ÷ 10   8   21 ÷ 7   36 ÷ 12   11 ÷ 11   90 ÷ 9

72 ÷ 9   2   12   18 ÷ 9   10   6

30 ÷ 10   54 ÷ 9   80 ÷ 10   10,000 ÷ 10   60 ÷ 12   5   70 ÷ 10   55 ÷ 11   5   7

63 ÷ 9   7   27 ÷ 9   54 ÷ 9   5   4

Division by 1 Digit (With Remainder)

302 r 1	242 r 3	9 r 2	468 ÷ 5	16 r 1
1,027 ÷ 3	115 ÷ 3 / 47 ÷ 2	629 ÷ 2	76 r 3 / 25 ÷ 4	34 ÷ 9 / 64 r 1	12 r 2
62 r 1		847 ÷ 2		93 r 3
848 ÷ 7	94 r 2	230 r 2	7 r 1	89 r 2
56 r 3	246 ÷ 5 / 28 ÷ 8	22 r 3 / 44 r 1	6 r 1	32 ÷ 7	314 r 1
42 r 2		1,939 ÷ 8	102 r 1	121 r 1
532 ÷ 9	373 ÷ 6	105 r 1		89 r 2
34 ÷ 3	133 ÷ 3 / 4 r 4	38 r 1	307 ÷ 4	78 r 5 / 32 ÷ 6	227 ÷ 4
269 ÷ 3			367 ÷ 3	97 ÷ 6
109 r 2	613 ÷ 6			338 ÷ 8
526 ÷ 5	3 r 7 / 56 ÷ 9	28 ÷ 3	10 r 1 / 629 ÷ 8	62 ÷ 5 / 17 ÷ 3	342 r 1
	692 ÷ 3	47 ÷ 5	59 r 1	566 ÷ 6
	122 r 1	423 r 1		76 r 4
5 r 2	1 r 6	11 r 1 / 3 r 4	41 ÷ 4 / 9 r 2	5 r 2
447 ÷ 5	384 ÷ 5	2,719 ÷ 9	29 ÷ 4	547 ÷ 5

Division by 2 Digits (With Remainder)

50 r 10	612 ÷ 30	159 r 3	2 r 56	55 r 7
767 ÷ 15 — 45 ÷ 2	8 r 1 — 7 r 55	15 r 8	368 ÷ 24 — 4 r 8	217 ÷ 37
20 r 12		180 ÷ 62		26 r 11
327 ÷ 20	1 r 24	987 ÷ 12	635 ÷ 24	462 ÷ 25
289 ÷ 32 — 49 ÷ 56	1 r 46	289 ÷ 36 — 156 ÷ 19	11 r 12 — 5 r 32	51 r 2
2,010 ÷ 40	5,149 ÷ 76	798 ÷ 5		870 ÷ 36
67 r 57		13 r 2		
22 r 1	277 ÷ 55 — 2 r 11	154 ÷ 25 — 9 r 1	4 r 1 — 5 r 2	59 ÷ 14 — 289 ÷ 72
34 r 24	186 ÷ 30	18 r 12	8,030 ÷ 80	19 r 24
1,248 ÷ 36	86 ÷ 31		18 r 5	24 r 6
447 ÷ 56	536 ÷ 82	87 ÷ 38	749 ÷ 67 — 250 ÷ 68	505 ÷ 63
	66 ÷ 42	3 r 5	82 r 3	
100 r 30	9 r 6		3 r 12	803 ÷ 41
96 ÷ 22 — 75 ÷ 12	6 r 3 — 6 r 44	8 r 1 — 4 r 3	3 r 46 — 6 r 9	4 r 8
210 ÷ 16	16 r 7	275 ÷ 15	997 ÷ 18	231 ÷ 73

Adding Fractions (Same Denominator)

Subtracting Fractions (Same Denominator)

Row 1 (reading cells):
- $2/10$ or $1/5$ — 0 — $4/9 - 3/9$
- $4/8$ or $1/2$ — $3/5 - 1/5$; $3/6 - 1/6$ — $5/6 - 4/6$
- $3/17 - 2/17$ — $4/8 - 2/8$
- $2/8$ or $1/4$ — $3/5 - 2/5$; $11/11 - 3/11$ — $3/5 - 2/5$
- $3/9 - 1/9$ — $6/10$ or $3/5$; $2/13 - 1/13$

Row 2:
- $8/7 - 3/7$ — $1/9$ — $1/8$
- $6/7 - 2/7$ — $4/7$ — $5/8 - 1/8$
- $1/13$ — $1/17$
- $4/8$ or $1/2$ — $2/5$ — $1/6$
- $4/5 - 1/5$ — $3/10 - 1/10$ — $2/4$ or $1/2$

Row 3:
- $5/8 - 1/8$ — $1/6$ — $1/5$
- $4/4 - 2/4$ — $2/6 - 1/8$ — $2/6 - 1/6$
- $2/6$ or $1/3$ — $1/12$
- $7/8 - 3/8$ — $3/6 - 1/6$
- $5/10$ or $1/2$ — $11/12 - 2/12$ — $3/11 - 1/11$
- $2/8$ or $1/4$ — $2/3 - 2/3$

Row 4:
- $6/8 - 2/8$ — $2/3$ — $1/3$
- $9/12$ or $3/4$ — $2/9$
- $4/8$ or $1/2$ — $2/8$ or $1/4$ — $8/11$
- $7/10 - 2/10$ — $5/7$
- $4/8$ or $1/2$ — $7/8 - 1/8$ — $2/11$; $3/10 - 1/10$
- $4/8$ or $1/2$

Row 5:
- $7/10 - 1/10$ — $2/6$ or $1/3$ — $3/5 - 1/5$
- $2/8$ or $1/4$ — $5/8$ — $2/11$ — $3/5$
- $6/8 - 2/8$ — $5/7$ — $3/11 - 1/11$
- $6/12 - 5/12$ — $1/5$ — $2/3 - 1/3$ — $3/8 - 1/8$
- $8/8 - 3/8$ — $3/8 - 1/8$ — $4/9 - 1/9$; $6/8$ or $3/4$ — $3/8 - 1/8$

Adding Fractions (Different Denominators)

$^3/_8 + ^1/_{16}$	$^5/_6 + ^1/_{12}$	$^4/_8$ or $^1/_2$	$^4/_{10}$ or $^2/_5$	$^{15}/_{16}$
$^1/_6 + ^1/_2$ / $^5/_{10}$ or $^1/_2$	$^1/_3 + ^2/_4$		$^2/_3 + ^1/_6$	$^{11}/_{12} + ^2/_6$ / $^{10}/_{10}$ or $\mathbf{1}$ / $^5/_8$
$^4/_{11} + ^1/_{22}$	$^2/_5 + ^1/_{10}$	$^2/_{17} + ^3/_{34}$		$^6/_{12} + ^1/_{24}$
$^{15}/_{16}$	$^7/_{34}$	$^5/_{10}$ or $^1/_2$ / $^2/_3 + ^1/_6$	$^5/_{10}$ or $^1/_2$	$^{10}/_{12}$ or $^5/_6$
$^5/_8$ / $^{10}/_{12}$ or $^5/_6$			$^5/_{12}$	$^1/_{10} + ^2/_5$ / $^3/_{12}$ or $^1/_4$
$^1/_5 + ^3/_{10}$	$^1/_5 + ^2/_{10}$	$^{15}/_{16}$		$^8/_{10} + ^2/_{20}$
$^5/_4$ or $\mathbf{1}$ $^1/_4$ / $^6/_8$ or $^3/_4$	$^2/_6 + ^1/_3$	$^5/_{10}$ or $^1/_2$ / $^2/_8 + ^2/_4$	$^4/_4 + ^1/_2$ / $^3/_{12} + ^1/_6$	$^{18}/_{20}$ or $^9/_{10}$ / $^{10}/_{12}$ or $^5/_6$ / $^4/_5 + ^2/_{10}$
$^3/_{10}$		$^5/_6$		$^1/_5 + ^1/_{10}$
	$^7/_{16}$		$^3/_4 + ^1/_{12}$	
$^{18}/_{22}$ or $^9/_{11}$	$^{16}/_{22}$ or $^8/_{11}$		$^5/_{10} + ^2/_5$	
$^1/_4 + ^7/_{12}$ / $^4/_6$ or $^2/_3$ / $^{13}/_{16}$		$^1/_6 + ^1/_3$	$^6/_4$ or $\mathbf{1}$ $^1/_2$ / $^1/_3 + ^1/_6$	$^4/_6$ or $^2/_3$
$^9/_{10}$	$^{11}/_{12}$	$^1/_4 + ^2/_8$	$^7/_{11} + ^2/_{22}$	$^7/_8 + ^1/_{16}$
$^{13}/_{24}$	$^9/_{22}$	$^{10}/_{12}$ or $^5/_6$	$^7/_8 + ^1/_{16}$	$^8/_{12}$ or $^2/_3$
$^1/_6 + ^1/_{12}$ / $^1/_4 + ^3/_8$	$^1/_4 + ^3/_8$ / $^5/_{16} + ^1/_2$	$^3/_6$ or $^1/_2$ / $^3/_6$ or $^1/_2$	$^1/_5 + ^3/_{10}$	$^{15}/_{12}$ or $\mathbf{1}$ $^1/_4$ / $^5/_6$
$^1/_2 + ^3/_4$	$^7/_8 + ^1/_{16}$	$^3/_6 + ^2/_{12}$	$^8/_{11} + ^2/_{22}$	$^1/_2 + ^4/_{12}$

Subtracting Fractions (Different Denominator)

$^4/_{12}$ **or** $^1/_3$ $^2/_5 - ^1/_{10}$ $^3/_{11} - ^2/_{22}$ $^{14}/_{20}$ **or** $^7/_{10}$ $^6/_{12} - ^1/_{24}$

$^3/_6$ **or** $^1/_2$ $^5/_{12} - ^2/_6$ $^1/_{12}$ $^5/_{16} - ^2/_8$ $^2/_{10}$ **or** $^1/_5$ $^7/_{10}$ $^3/_6 - ^1/_{12}$ $^1/_8$

$^1/_2 - ^4/_{12}$ $^5/_6 - ^1/_{12}$ $^6/_8 - ^1/_4$

0 $^3/_6$ **or** $^1/_2$ $^2/_8$ **or** $^1/_4$ $^5/_{10}$ **or** $^1/_2$ $^4/_8$ **or** $^1/_2$ $^1/_{34}$ $^7/_8 - ^1/_{16}$

$^3/_8 - ^1/_{16}$ $^3/_{12} - ^1/_6$ $^4/_5 - ^1/_{10}$ $^6/_8 - ^2/_4$ $^2/_{10}$ **or** $^1/_5$ $^1/_{12}$ $^3/_8$

$^{11}/_{20}$ $^{11}/_{24}$

$^7/_{22}$ $^4/_5 - ^5/_{20}$ $^{12}/_{22}$ **or** $^6/_{11}$

$^3/_4 - ^3/_8$ $^1/_{16}$ $^2/_5 - ^2/_{10}$ $^3/_6$ **or** $^1/_2$ $^4/_4 - ^1/_2$ $^7/_{12} - ^1/_4$ $^4/_{12}$ **or** $^1/_3$ $^2/_6 - ^1/_3$

$^{13}/_{16}$ 0 $^3/_4 - ^1/_{12}$ $^1/_{10}$

$^2/_{12}$ **or** $^1/_6$ $^7/_8 - ^1/_{16}$ $^9/_{12}$ **or** $^3/_4$

$^2/_4$ **or** $^1/_2$ $^6/_{10} - ^2/_5$ $^5/_{12}$ $^2/_3 - ^1/_6$ $^3/_8 - ^1/_4$ $^3/_3 - ^2/_4$ $^6/_{12}$ **or** $^1/_2$

$^7/_{11} - ^2/_{22}$ $^8/_{10} - ^2/_{20}$ $^7/_8 - ^1/_{16}$ $^4/_5 - ^3/_{10}$ $^3/_{10}$

$^5/_{10} - ^2/_5$ $^{13}/_{16}$ $^8/_{12}$ **or** $^2/_3$ $^1/_4 - ^2/_8$ $^5/_{16}$

$^2/_3 - ^1/_6$ $^4/_6 - ^1/_2$ $^3/_{10}$ $^5/_6 - ^1/_3$ $^3/_6$ **or** $^1/_2$ $^2/_3 - ^1/_6$ $^1/_6$ $^3/_5 - ^3/_{10}$

$^{13}/_{16}$ $^4/_{22}$ **or** $^2/_{11}$ $^3/_6 - ^2/_{12}$ $^2/_{17} - ^3/_{34}$ $^4/_{11} - ^1/_{22}$

Reducing Fractions to Simplest Form

9/16	2/5	3/4	20/50	5/6
12/14 ... 1/2 ... 3/18	3/4 ... 100/100	12/16	1/4	3/4
25/30			12/16	3/18
1/2	2/3	7/12	10/46	
10/24 ... 1/6	5/12 ... 2/8	1/2 ... 25/75	25/100 ... 2/3	3/5
	22/24	3/8	15/30	
1/10	2/3	1/9	1/3	
1/2 ... 3/12	1/4	12/16	1 ... 15/25	10/40
6/16	4/6	9/30		20/50
3/8		3/10		
12/18	1/4 ... 5/10	3/4 ... 7/14	6/8	1/4
6/9	2/3	10/16	2/5	14/24
3/10	1/4	200/300	5/8	11/12
2/10 ... 7/9 ... 4/8	1/5	1/5 ... 1/8	1/3 ... 2/10	10/80
2/6	5/23	100/1000	6/20	3/12

Adding Decimals

16.26 $0.42 + $0.03 24.50 + 1.79 95.1	8.7 207.40 + 153.20 7.12 + 0.58	$9.40 2.3	eighty-five cents 4.85 1.25 + 0.075	5.29 99.80 + 78.19 9.89 + 7.45 $15 + $5.25
0.60 + 0.20 123.41 + 1.4 32.95 8.75 + 6.25	0.80 $1.10 + $0.75	0.25 + 0.14 7.7 7.9 4.75 + 4.05	20.25 104.20	9.210 + 3.0 17.34 11.99 3.0 + 0.9
8.8 1.2 + 1.1 $1.50 75.3 + 1.4 15.80 + 1.70	235.73 4.7 + 0.5 234.60 + 1.13	0.55 9.66 7.45 + 1.25	$1.85 $2.21 + $1.25 one $0.45 + $0.17	
3.4 $1.29 + $0.75 2.71 + 2.58	forty-five cents 17.50 26.29 6.54 + 3.35	1.33 $1.10 + $0.40 360.6	124.81 $10 + $0.80	9.89 + 7.25 4.64 + 0.21 177.99 $68.24 + $68.20
3.9 78.5 + 25.70 8.99 + 0.67 1.29 + 0.04	$136.44 5.25 + 4.9 4.9 + 3.0 $4.85 + $4.55	$10.80 $3.46 2.17	31.65 + 1.30 11.5	1.12 + 1.05 5.2 $86.45 + $7.20 $0.75 + $0.10
6.5 + 5.0 7.97 $3.25 8.66 + 7.60	15 0.98 + 0.02 2.6 + 0.8	sixty-two cents $2.50 + $0.75 $93.65 17.14	9.89 $2.04 11.11 + 0.88 0.31 + 0.24	1.325 10.15 0.39

Subtracting Decimals

1.8	6.20	14.10	sixty-five	2.1
$1.29 - $0.75 / $0.42 - $0.05	54.20	22.71	78.5 - 25.70 / 4.64 - 3.21	8.99 - 0.67
2.71 - 2.58		3.19	1.25 - 0.75	1.29 - 0.04
6.54 - 3.35	0.07	31.65 - 1.30		$9.75
11.11 - 0.88	8.32	4.7 - 0.5	122.01 / 4.9 - 3.0	52.80
0.31 - 0.24	7.45 - 1.25	1.5	$10 - $0.80	4.75 - 4.05
89.1	233.47	0.11	1.06	0.7
2.44 / 10.23	75.3 - 1.4	6.54	24.50 - 1.79 / $0.39	0.1 / $0.70
3.0 - 0.9	15.80 - 1.70		92.10 - 3.0	
1.25	2.64	0.13		2.5
$1.10 - $0.40	207.40 - 153.20 / 1.43	96.80 - 78.19 / 21.61	123.41 - 1.4 / 9.89 - 7.45	$0.98 - $0.09 / 0.40
	$68.24 - $64.24	$15 - $5.25	$1.10 - $0.75	$1.75 / 2.6 - 0.8
0.5	four dollars		1.12 - 1.05	thirty cents
0.35	5.25 - 4.9 / 1.9	4.2	$79.25 / 7.12 - 0.58	1.2 - 1.1
0.25 - 0.14	$4.85 - $4.55	234.60 - 1.13	0.75 - 0.10	
6.5 - 5.0		thirty-five cents	$9.20	0.28
$2.50 - $0.75 / 73.9	0.60 - 0.20 / 30.35	$2.21 - $1.25 / $0.89	$0.96	$86.45 - $7.20 / fifty-four cents
8.66 - 7.60	8.75 - 6.25	0.45 - 0.17	0.07	9.89 - 7.25

Multiplying Decimals

3.0

2.2 × 2

5 × 1.1

$2.12

6.66

2.3 × 3

0.30

10.4

4.4

6.24

6.16

1.4 × 2

7.5 × 5

16.4

0.10 × 3

9.1 × 3

4.2 × 3

21.3

0

0.3 × 2

4.1 × 1

1.35

8.02 × 4

$1.06 × 2

8.2 × 3

3.33

1.2 × 3

1.11 × 4

6 × 2.1

12.6

5.5

90.9

32.08

1.3 × 2

4.2

5.2 × 2

15.6

5.2 × 3

13.6 × 2

7.1 × 3

1.2

1.4 × 3

37.5

7.01 × 3

6.9

27.3

18.8

0.6

4.44

2.8

3.08 × 2

2.5

24.6

21.03

1.04 × 4

9.4 × 2

11.7 × 3

10.1 × 9

2.22 × 3

4.16

0 × 7.1

3.6

2.6

25

0.3 × 4

14.2

7.1 × 2

4.8

4.4

0 × 12.8

12.6

27.2

12.5 × 2

8.2 × 2

0.5 × 5

1.11 × 3

0.5 × 6

1.04 × 6

Fraction and Decimal Conversion

twenty-nine cents | 1.6 | nine and 2 tenths | $^1/_3$ | 0.6

$^4/_4$ | $^3/_{10}$ | $^3/_{10}$ | $^4/_{100}$ | $^1/_2$ | 10 cents | 0.80 | $^3/_4$

9 dimes | $^1/_5$ of $1.00 | | fifty cents | $^3/_2$

0.12 | 0.01 | 0.04 | 1.5

0.33 | 0.75 | 0.33 | $^1/_4$ | $^1/_2$ | $^1/_2$ | $^{50}/_{100}$ | 9.2 | $^6/_{10}$ | $^{100}/_{100}$

$0.29 | 0.09 | | $^4/_2$

$0.80 | $0.12 | 0.18 | $0.50

0.10 | 1 | $^{77}/_{100}$ | $^{75}/_{100}$ | 1 | $^{75}/_{100}$ | $^5/_{10}$

four hundredths | 5 nickels | one dollar | 2 dimes, 1 penny

twenty cents | $1.00 | $0.50 | sixty-eight hundredths

$^2/_2$ | $0.75 | 0.75 | $^{33}/_{100}$ | 1.00 | $^1/_3$ | 9.0 | $^3/_4$

twelve cents | twenty-five cents | 80 cents | $^2/_6$ | $^{70}/_{100}$

2.0 | $0.11 | 0.70

0.25 | 0.50 | $^1/_{10}$ of $1.00 | $^{10}/_{100}$ | $^1/_2$ | 0.04 | 0.7

one and six tenths | one hundredth | eleven cents | 1/5 | 12 pennies

$0.25 | $0.12 | $0.25 | 0.20 | nine hundredths

0.77 | $^{25}/_{100}$ | $^2/_4$ | $^{80}/_{100}$ | 75 cents | 0.25 | 0.3

six tenths | | 2 quarters | $^{68}/_{100}$ | $^{81}/_{100}$

Comparing Decimals (Greater Than, Less Than, and Equal To)

greater than
5.3 ? 5.6
9.0 ? 9.5

less than
=
5.7 ? 3.6

greater than
7.1 ? 0.3

less than
greater than
3.2 ? 0.9

less than
greater than
$0.75 ? $0.25
greater than

greater than
1.25 ? 1.40

less than
greater than
5.7 ? 3.8

greater than
=
=
0.5 ? 5.0

less than
greater than
8.7 ? 8.6
2/10 ? 0.2
10.1 ? 1.1

less than
greater than

greater than
0.53 ? 0.47

less than
9.2 ? 6.4
6.4 ? 2.8

less than
less than
4.1 ? 4.01
0.4 ? 0.2
4.2 ? 3.9
0.2 ? 2.0

greater than
0.7 ? 0.6
0.9 ? 0.3

greater than
6.4 ? 6.8
greater than

greater than
1.3 ? 1.2
8.2 ? 8.8

less than
7.17 ? 7.56
3.6 ? 3.8

less than
5.3 ? 5.2
$0.50 ? $0.75
4.3 ? 4.5
greater than

greater than
$7.63 ? $7.22
greater than

less than
0.34 ? 0.25
4.8 ? 4.5

greater than
greater than
1 ½ ? 1.5
1.0 ? 0.6

greater than
3.9 ? 4.8

greater than
$31.75 ? $31.25
greater than

less than
$2.50 ? $1.62
11.2 ? 11.0

0.60 ? 0.65
=
4.5 ? 3.7

greater than
4.4 ? 4.5
8.2 ? 1.0

less than
18.6 ? 18.2

greater than
3.5 ? 3.50
=

greater than
7/100 ? 0.07
23/100 ? 0.23
2.0 ? 0.2

greater than
1/10 ? 0.10
=
1.3 ? 1.4

Basic Algebra

$60n = 1{,}200$ | $n = 10$ | $50 - x = 25$

$40n = 240$ | $n = 9$ | $n = 11$ | $x = 4$

$72 = n + 63$ | $n = 9$ | $4 + 9 + 3 + n = 20$

$x = 100$ | $12n = 120$ | $4 \times n = 400$ | $8n = 72$

$129/n = 129$ | $x = 25$ | $4 \times n = 20$

$3n = 21$ | $8 \times n = 56$ | $2x = 16$ | $n + 15 = 25$

$120 - n = 80$ | $4x = 28$ | $n = 5$ | $n = 8$

$8 = n$ | $9n = 81$ | $30 - x = 10$ | $7 = n + 12$

$5n = 50$ | $2x = 10$ | $36 - n = 6$

$n = 4$ | $n = 2$ | $10x = 1{,}000$ | $7 = n$ | $x = 10$

$n = 33$ | $x = 7$ | $x = 20$ | $3n = 30$

$n = 10$ | $7 + n = 15$ | $n = 9$ | $x = 12$

$n = 100$ | $5 + 6 + 7 + n = 20$

$n + 24 = 32$ | $n = 150$ | $3x = 24$ | $7x = 56$

$n = 10$ | $n = 10$ | $0 = n$ | $x = 5$

$n = 7$ | $n = 40$ | $4 + 8 + n = 45$ | $8 = x$

$3x = 36$ | $n = 20$

$x = 8$ | $x = 8$ | $10 \times n = 100$ | $n = 1$

$n = 11$ | $25x = 100$ | $350 - n = 200$

$n + 8 = 16$ | $4 = 15 - n$ | $30 = 19 + n$

$n + 3 + 5 = 15$ | $n = 5$ | $8 = n$

$n + 43 = 43$ | $n = 30$ | $n = 7$ | $8x = 80$

Algebra: Addition and Subtraction

Row 1

- $3 + 4 - n = 6$ | $x = 0$ | $n = 6$ | $10 + 10 - n = 6$
- $n = 3$ | $20 - 10 + n = 13$
- $5 - 4 + x = 12$ | $7 + n - 3 = 10$ | $50 - x + 40 = 80$
- $x = 10$ | $n = 7$ | $10 + 8 - n = 15$
- $n = 25$ | $n + 6 - 5 = 10$ | $x = 2$ | $300 + 200 - x = 100$

Row 2

- $x = 4$ | $4 - x + 3 = 6$ | $n = 1$
- $15 - n + 10 = 20$ | $n + 4 - 3 = 11$ | $n + 20 - 3 = 25$
- $8 + n - 1 = 21$ | $2 + 8 - n = 5$
- $n = 5$ | $16 - n + 0 = 8$ | $7 + 7 - n = 14$ | $75 - 25 + n = 60$
- $32 = 25 + 15 - n$ | $x = 50$ | $x = 1$ | $80 = 50 + 40 - n$

Row 3

- $n = 30$ | $x - 3 + 1 = 10$
- $n = 12$ | $n - 2 + 5 = 15$ | $7 + 3 - n = 8$ | $20 - n + 3 = 10$
- $n = 14$ | $n + 3 - 5 = 10$
- $16 = x + 20 - 8$ | $n = 5$ | $21 = 15 + 15 - n$
- $n = 10$ | $x - 500 + 0 = 500$ | $3 + 7 + x = 10$ | $150 + 50 - n = 175$

Row 4

- $n = 5$ | $20 - x + 4 = 15$ | $45 - 10 + n = 50$
- $n = 0$ | $x = 4$ | $x = 4$
- $n = 15$ | $x = 7$ | $15 - 5 + n = 21$
- $n = 11$ | $x = 12$
- $x = 5$ | $21 = 14 + x - 0$ | $n = 9$ | $45 - 15 + n = 60$

Row 5

- $x = 3$ | $x = 9$ | $30 + 20 - x = 45$
- $x = 400$ | $x = 1{,}000$ | $n = 12$
- $n = 13$ | $x = 3$ | $n = 10$ | $x = 11$
- $n = 10$ | $n = 11$ | $40 - 30 + n = 15$
- $n = 6$ | $n - 6 + 5 = 10$ | $100 - x + 0 = 50$ | $24 - x + 4 = 25$

Row 6

- $n = 8$ | $10 + 10 - n = 19$ | $n = 2$ | $17 = 14 - x + 7$
- $n = 6$ | $x = 100$ | $n = 8$
- $14 - x + 10 = 20$ | $n = 1$ | $n = 8$ | $18 - n + 6 = 18$
- $n = 250$ | $x - 50 + 25 = 75$ | $7 - x + 4 = 8$ | $n = 3$
- $n = 14$ | $x + 8 - 5 = 5$ | $15 - 8 + n = 14$ | $450 + 50 - n = 250$

Squares

A cut-apart matching puzzle grid (5 × 5 pieces). Each piece has a squared expression or value on each edge. Some text appears rotated/upside-down.

40^2 · 18^2 · 30^2	50^2 · 75^2 · 144 · 80^2	$4{,}900$ · 16^2 · 12^2 · $6{,}400$	900 · 11^2 · 100	40^2 · 21^2 · $1{,}000{,}000$
36 · 100 · 90^2	80^2 · 196 · 25	1 · 441 · 324 · 9 · $2{,}500$	1 · 15^2 · 36 · 400	$1{,}000{,}000$ · 20^2 · 81
$8{,}100$ · 225 · 70^2	25 · 14^2 · 361	25^2 · 400 · 289 · $1{,}600$	36 · 17^2 · 13^2 · 1^2	25 · 16 · $10{,}000$
2^2 · 6^2 · 60^2	9^2 · 256 · 5^2	121 · 6^2	5^2 · 1^2 · 5^2 · 4	20^2 · 7^2 · 36 · 6^2
$3{,}600$ · 169 · $1{,}600$	4^2 · 7^2	10^2 · 3^2 · 19^2	$5{,}625$ · 8^2 · 49 · 625	100^2 · 10^2 · 49 · 64 · $1{,}000^2$

Cubes

50³	4³	729	8,000	2³
20³ · 2,744	3,375 · 15,625	12³ · 125	6³	18³
64	50³	2³	80³	
27,000	8	42,875	1,000,000,000	
25³	14³	6³	30³ · 216 · 7³	2,197
30³	5³	27	5³	
125,000	3³	343,000	60³	
1³ · 27	10³ · 13³	5³ · 1,331	8	
35³	1³	125,000	70³	
125	512,000	216,000		
5,832 · 8,000	512 · 1,000 · 4³ · 8³ · 216 · 64	15³		
60³	9³	343		
125	1,000,000	1	7³	
27 · 11³ · 27,000 · 3³ · 1 · 343 · 2³ · 1,728 · 3³				
1,000³	216,000	8	20³	100³

Square Roots

200	√1,600	60	√22,500	√289
√36 · 16	√49	√81 · √1,000,000 · 70	0 · √169	√144 · √900
√8,100	√4,900		120	√2,500
9	6	3 · 5	11 · √0	√225 · √196
√576 · 75	21	√1,225 · 110	√4	90 · 25
4	1,000	20 · √361	12	15
√100	√625	40	√12,100	√3,600 · 30
2	6 · 17	10 · √441	√25 · √121 · √324	√36 · √64 · √529 · 19
23	24	18	35	√484
√256	√9 · 7	√16 · 100	13 · √400	√10,000 · 14 · 8
150		22	50	√40,000

Solution: Two Addends

23 (12+11)	86 (21+10)	278 (5+16)	243 (32)	268 (8+15)
48 (72+13)	8+12 (12+15)	486 (28)	173 (12+6)	92 (18)
216 (100+21)	21 (16+8)	285+4 (18+10)	224 (14+7)	413 (21)
17+9 (32+45)	30 (21)	36 (9+12)	215 (15+6)	82 (32+13)
26 (41+54)	20+10 (15+10)	25+11 (25)	200+15 (44+11)	47+35 (12+11)

Solution: Three Addends

53 (36+24+1)	56+78+125 (190)	$27 (28+78+84)	16+13+78 (83+34+17)	12+23+56 (14+25+121)	76+78+34 (396+87+9)
17+8+34 (25)	23+80+56 (33)	25+16+78 (74)	99+56+44 (62+25+16)	269 (103)	47+34+98 (96)
213 (152)	5+56+13 (120)	98+160+34 (70)	396 (126)	56+77+10 (87)	77+90+21 (39)
87+17+245 (81)	777+888+3 (70)	676+34+17 (157)	45+87+45 (75)	33+21+22 (112)	76 (123)
(111+22+8) 141	(57+89+33) 176	(43+17+41) 101	(11+2+11) 40	(89+21+13)	

38 Puzzled by Math!

© 2005 Trish Caldwell Landsittel

Solution: Addition: Mental Math

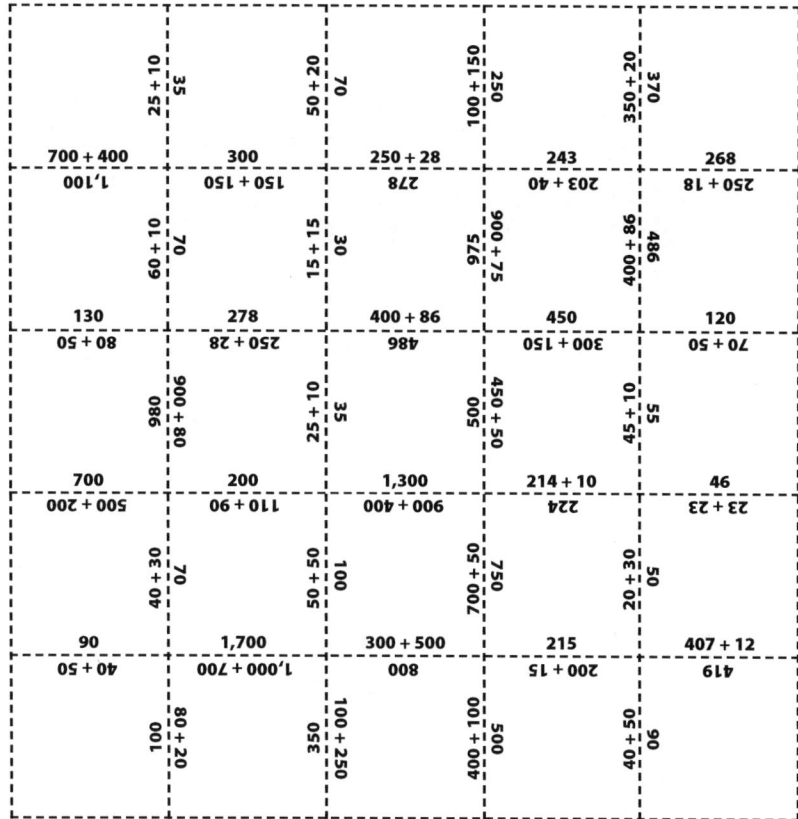

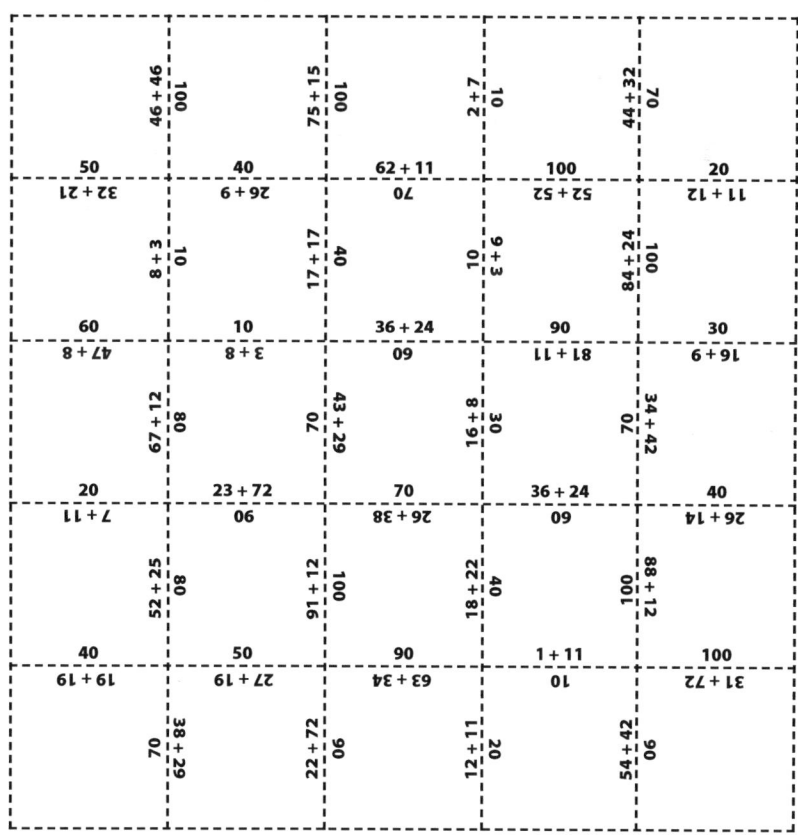

Solution: Estimate the Sum

Puzzled by Math!

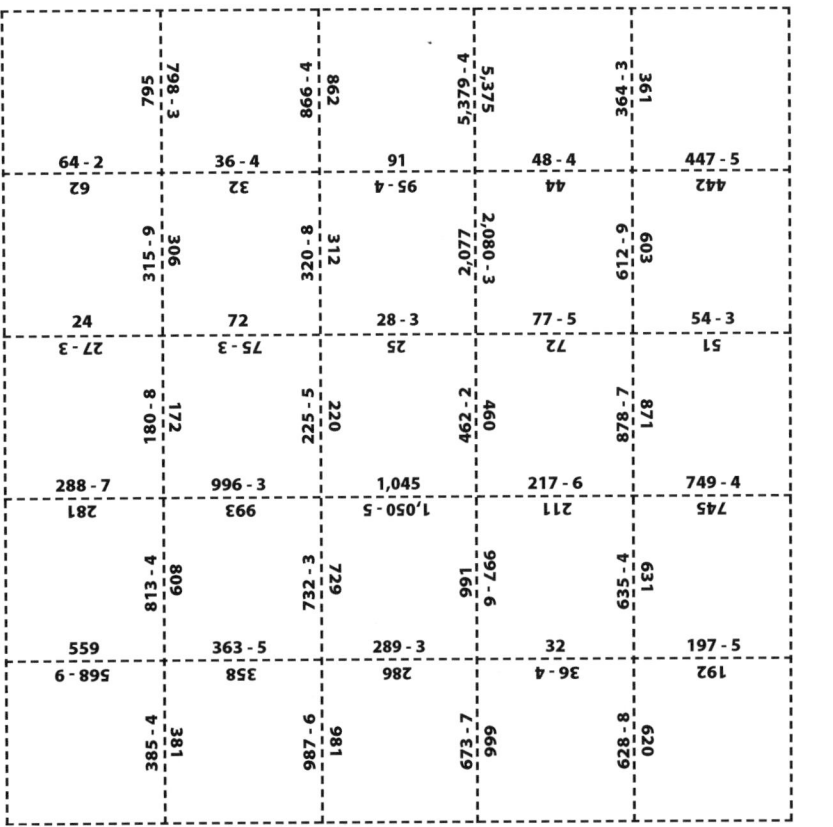

Solution: Subtracting by 1 Digit

Solution: Subtracting by 2 Digits

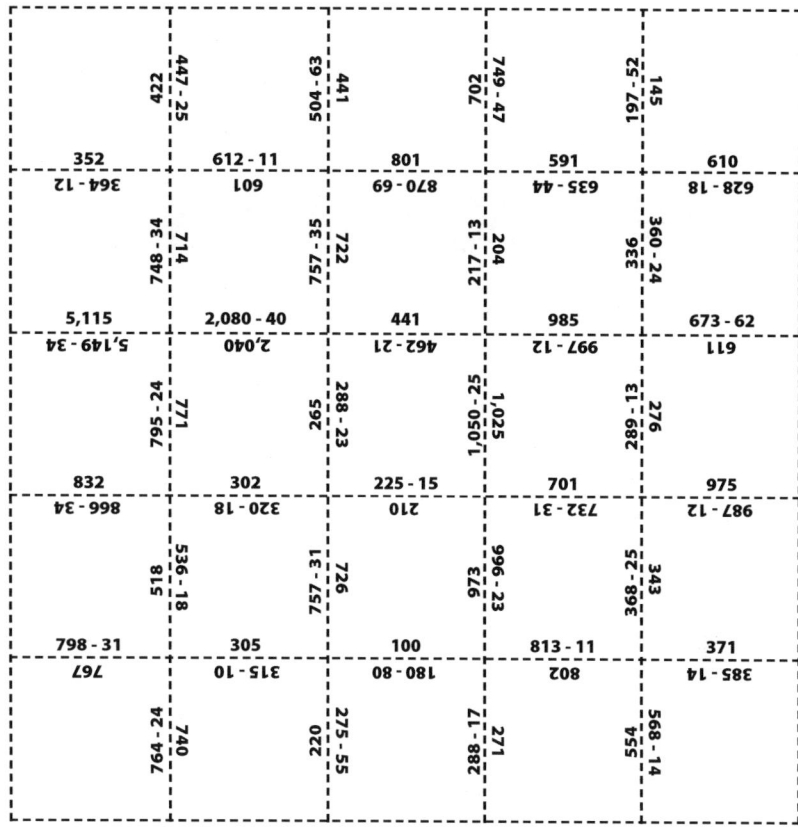

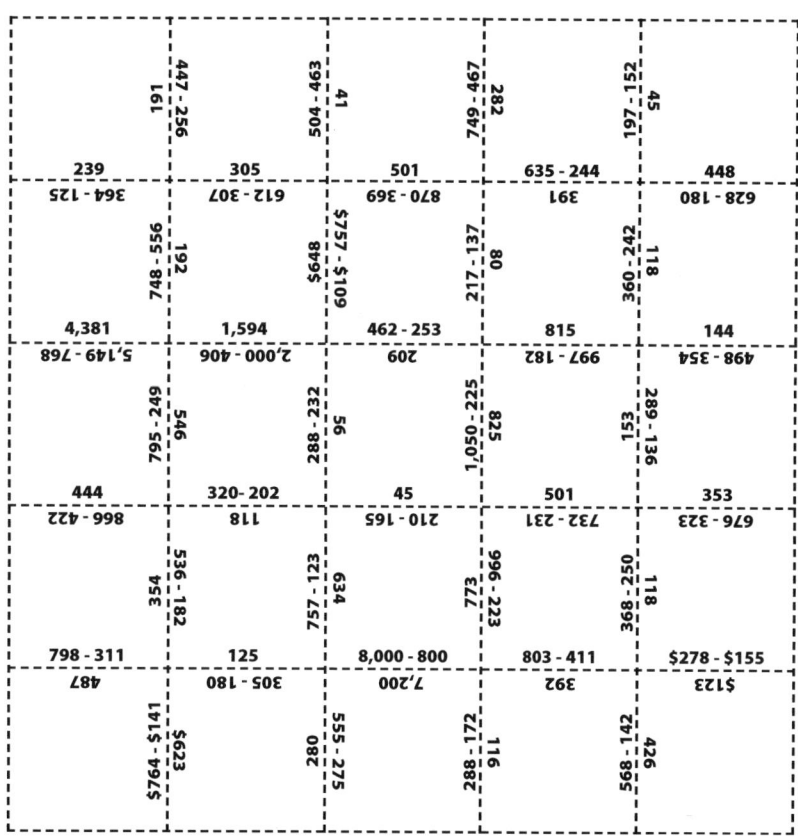

Solution: Subtracting by 3 Digits

Solution: Estimate the Difference

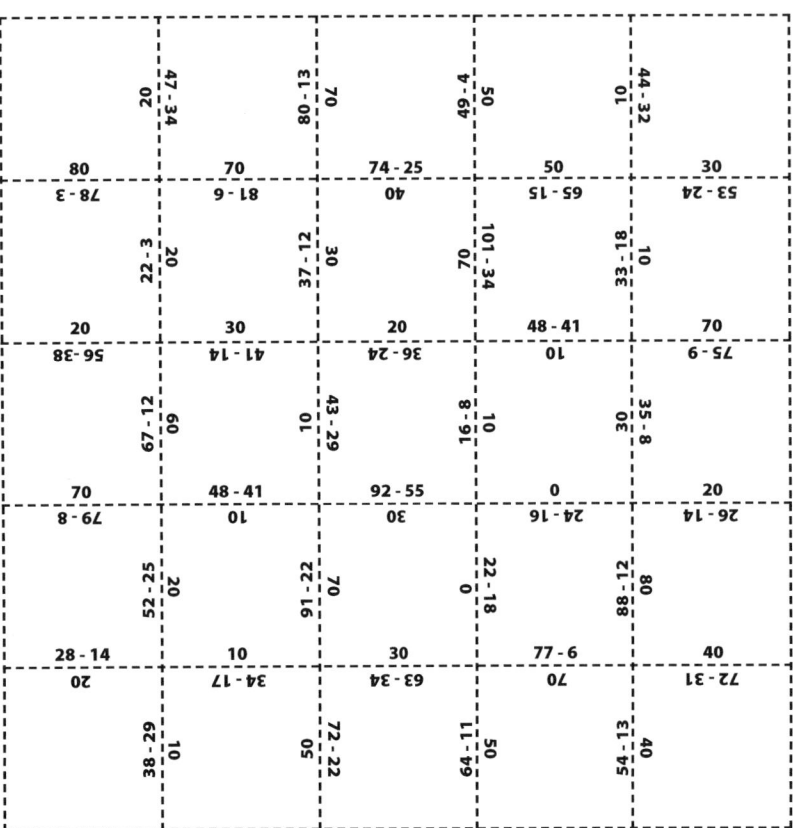

Solution: Subtraction Practice

Answer Key

Solution: Multiplication by 1s, 2s, 3s, & 4s

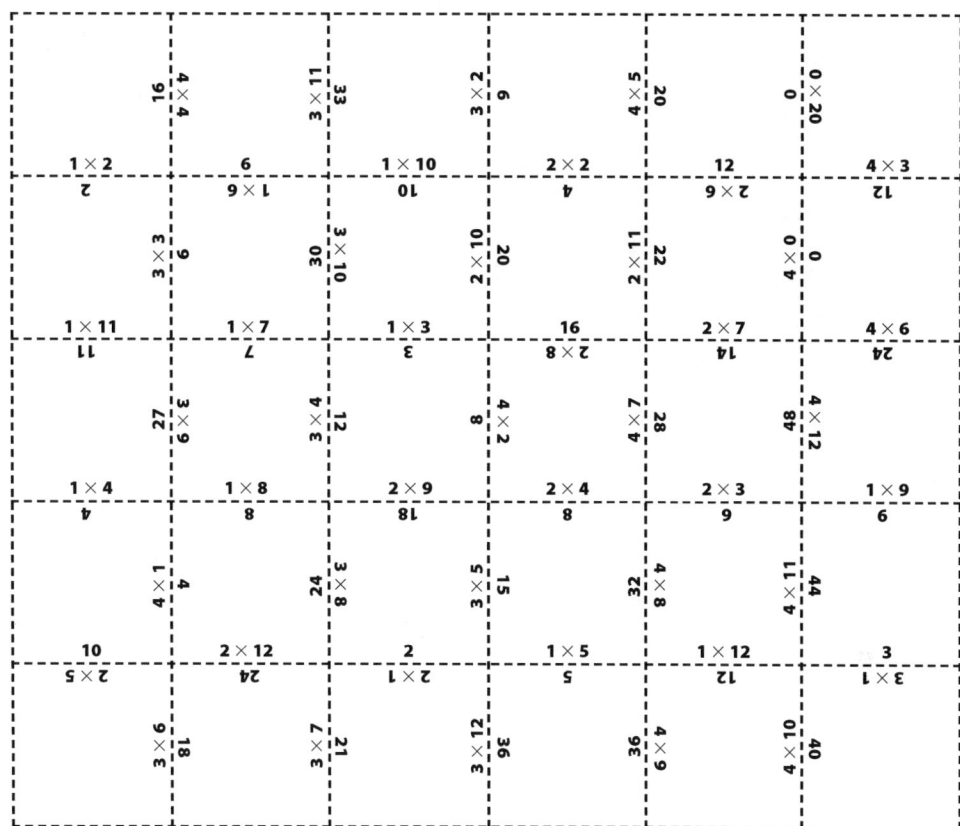

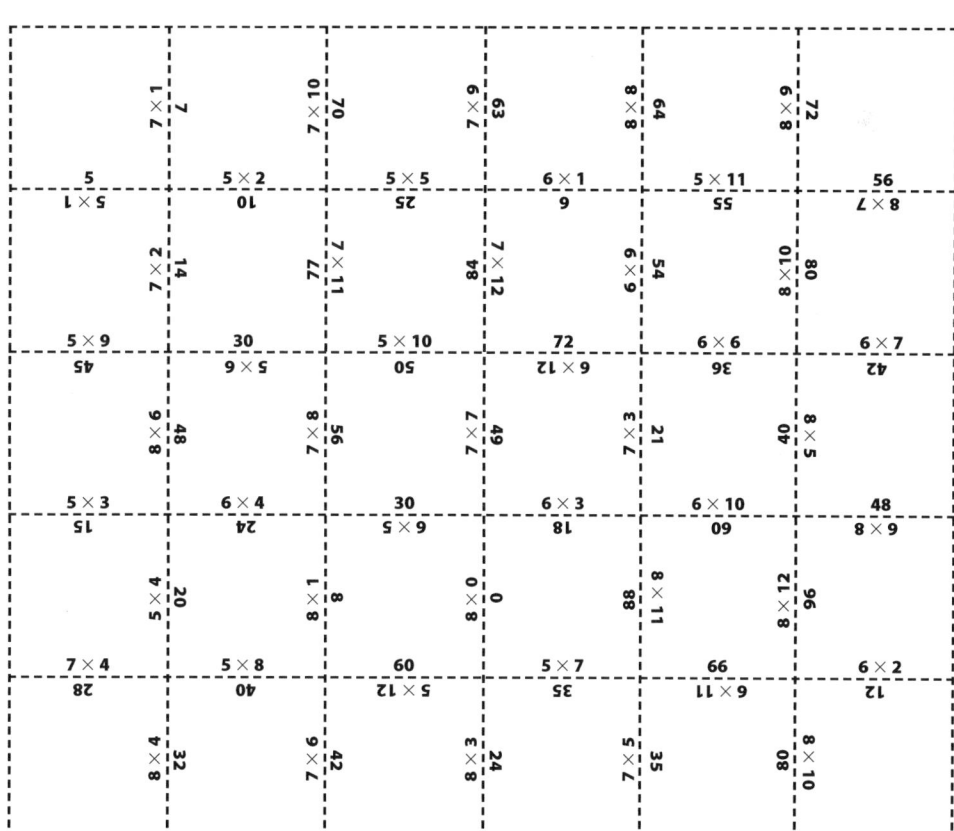

Solution: Multiplication by 5s, 6s, 7s, & 8s

Solution: Multiplication by 9s, 10s, 11s, & 12s

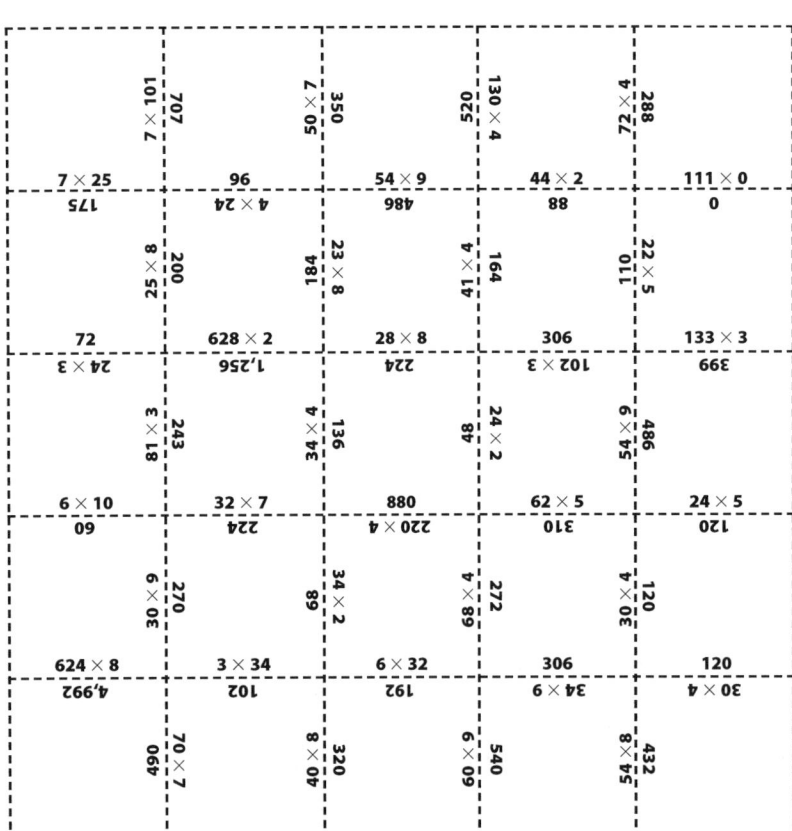

Solution: Multiplication by 1 Digit

Solution: Multiplication by 2 Digits

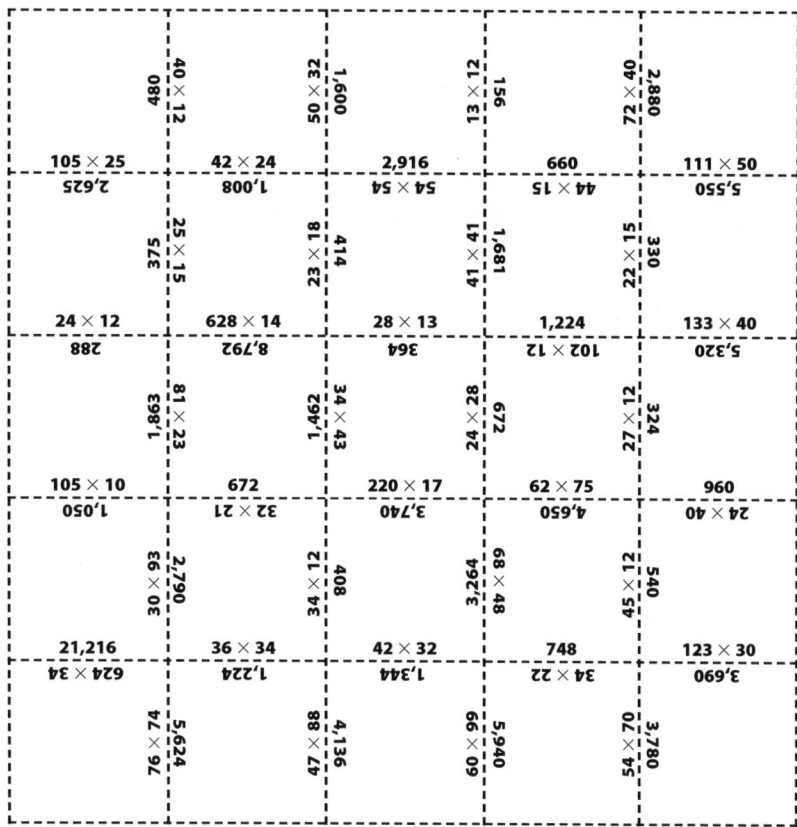

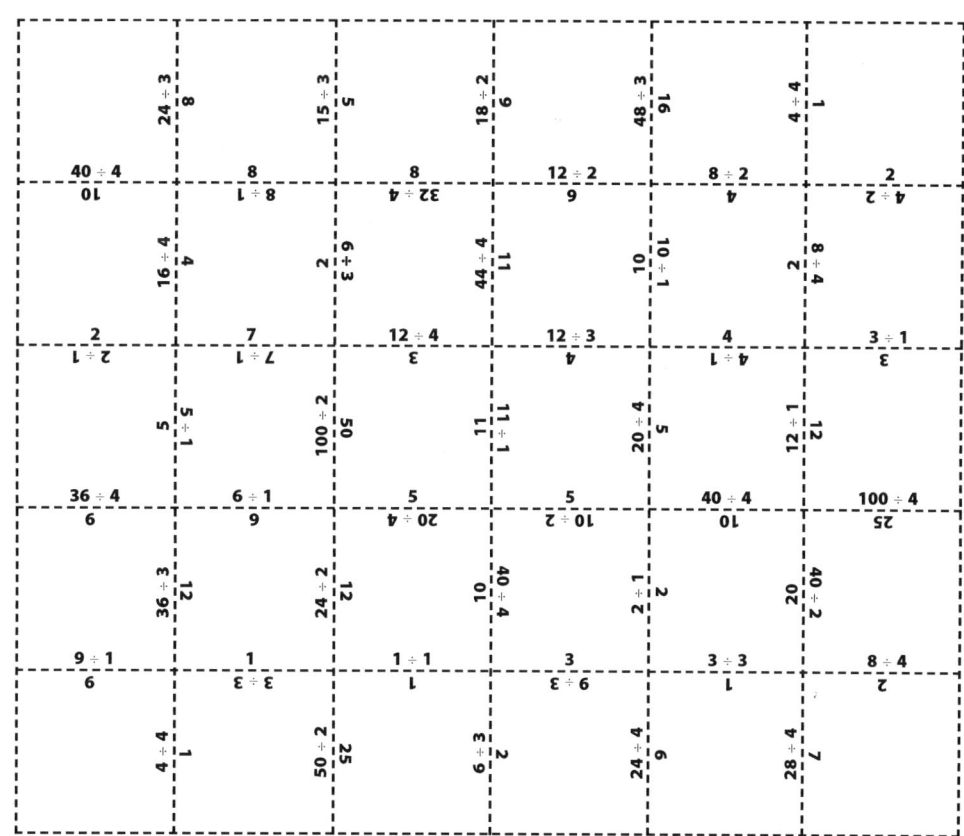

Solution: Division by 1s, 2s, 3s, & 4s

Solution: Division by 5s, 6s, 7s, & 8s

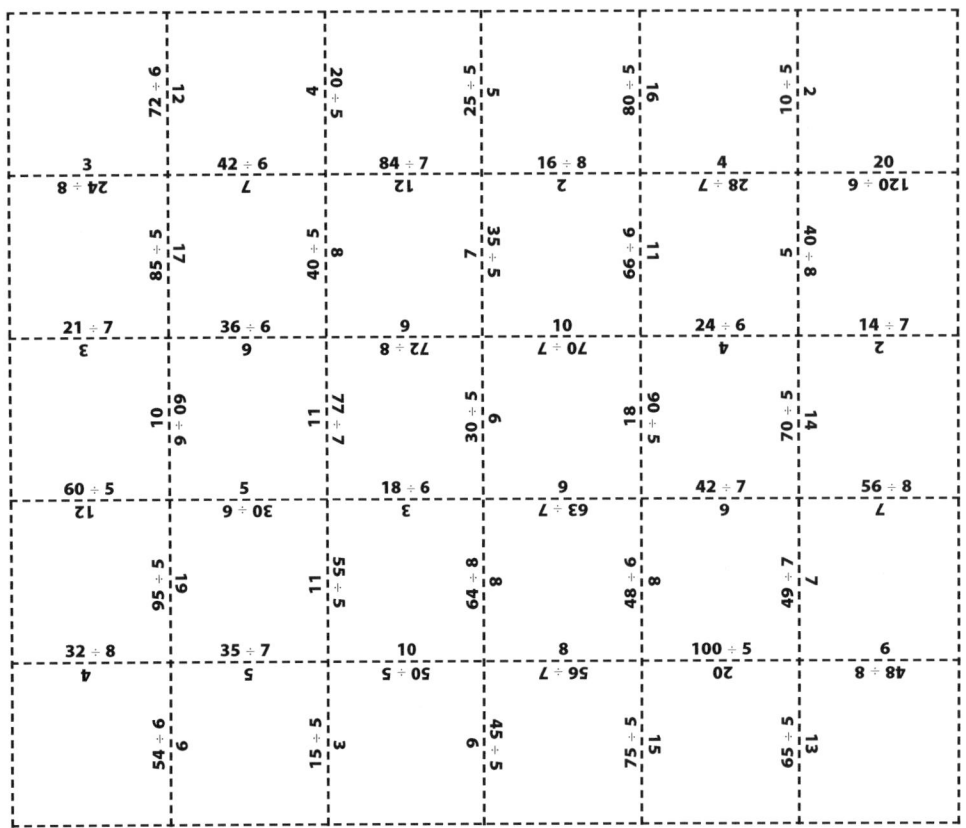

Solution: Division by 9s, 10s, 11s, & 12s

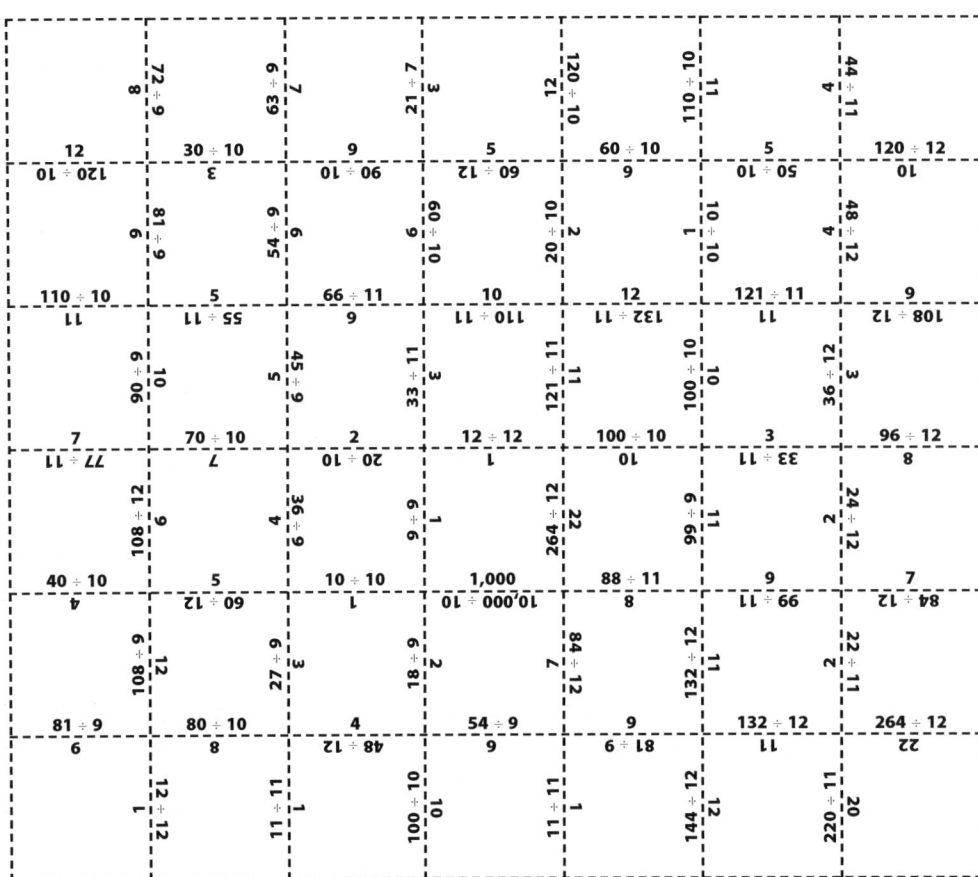

Solution: Division by 1 Digit (With Remainder)

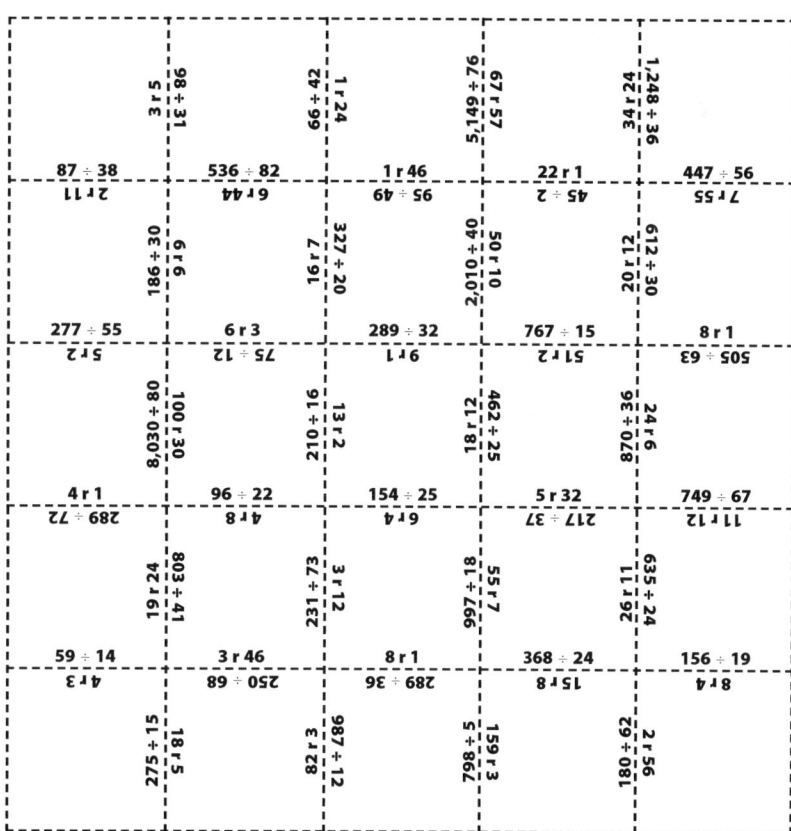

Solution: Division by 2 Digits (With Remainder)

Solution: Adding Fractions (Same Denominator)

3/6 or 1/2 / 2/6 + 1/6	1/5 + 3/5 / 5/5	5/6 / 1/6 + 4/6	6/8 or 3/4 / 5/8 + 1/8
11/12	5/8 + 1/8	4/5	5/7	3/6 + 1/6
4/6 or 2/3 / 6/12 + 5/12	8/8 or 1 / 6/10 + 3/10	3/5 + 1/5	3/7 + 2/7 / 3/8	4/6 or 2/3 / 2/8 + 1/8
4/11 / 1/11 + 3/11	7/8 + 1/8	2/3	1/7 + 3/7	3/4
6/8 + 2/8	8/8 or 1	1/3 + 1/3	4/7	1/4 + 2/4
7/7 or 1 / 6/7 + 1/7	4/8 or 1/2 / 3/8 + 1/8	3/11 + 4/11 / 2/11	2/2 or 1 / 1/2 + 1/2	
9/10	3/8 + 1/8	11/12	2/8 + 3/8	1/10 + 1/10
4/6 or 2/3 / 3/6 + 1/6	4/8 or 1/2 / 7/10 + 2/10	1/5 + 2/5 / 5/5	1/3 + 1/3 / 2/3	4/8 or 1/2 / 3/8 + 1/8 / 2/10 or 1/5
4/8 + 3/8	7/10 + 1/10	1/8 + 2/8	3/5	1/11 + 3/11
7/8 / 2/5 / 1/5 + 1/5	8/10 or 4/5 / 3/13 / 2/13 + 1/13	3/8 / 1/17 + 3/17 / 4/17	1/5 + 2/5 / 7/10 / 3/10 + 4/10	4/11

1/6 / 2/6 - 1/6	3/5 - 2/5 / 1/5	5/6 - 4/6 / 1/6	4/8 or 1/2 / 5/8 - 1/8
1/12	5/8 - 1/8	2/5	4/7	3/6 - 1/6
2/11 / 3/11 - 1/11	4/8 or 1/2 / 7/8 - 3/8 / 3/10 - 1/10	2/10 or 1/5 / 4/9 - 3/9 / 1/9	6/7 - 2/7 / 1/8	2/6 or 1/3 / 2/8 - 1/8
6/8 - 2/8	7/8 - 1/8	0	8/7 - 3/7	4/4 - 2/4
2/4 or 1/2 / 6/7 - 1/7 / 5/7	6/8 or 3/4 / 2/8 or 1/4 / 3/8 - 1/8	2/3 - 2/3 / 2/11 / 3/11 - 1/11	5/7 / 4/5 - 1/5 / 3/5	2/4 or 1/2
7/10 - 2/10	3/8 - 1/8	11/12 - 2/12	5/8	3/10 - 1/10
5/10 or 1/2 / 2/6 or 1/3 / 3/6 - 1/6	2/8 or 1/4 / 2/5 / 3/5 - 1/5	9/12 or 3/4 / 8/8 - 3/8 / 1/3	2/3 - 1/3 / 1/3	2/10 or 1/5 / 2/8 or 1/4 / 3/8 - 1/8
7/8 - 3/8	7/10 - 1/10	6/8 - 2/8	1/5	11/11 - 3/11
4/8 or 1/2 / 2/9 / 3/9 - 1/9	6/10 or 3/5 / 1/13 / 2/13 - 1/13	4/8 or 1/2 / 3/5 - 2/5 / 1/17 - 3/17	3/5 - 2/5 / 3/17 - 2/17 / 1/17	2/8 or 1/4 / 4/8 - 2/8 / 8/11

Solution: Subracting Fractions (Same Denominator)

Solution: Adding Fractions (Different Denominator)

Solution: Subtracting Fractions (Different Denominator)

Solution: Reducing Fractions to Simplest Form

4/6 2/3	9/30 3/10	20/50 2/5		12/16 3/4
3/12 1/4	1/4 25/100	7/14 1/2	12/16 3/4	100/100 1
14/24 7/12	6/16 3/8	5/6 25/30	3/18 1/6	
6/8 3/4	25/75 1/3	12/14 6/7	1/4 10/40	12/16 3/4
10/16 5/8	9/30 3/10	1/3 2/6	2/5 20/50	
5/10 1/2	1/8 10/80	2/10 1/5	15/25 3/5	3/18 1/6
22/24 11/12	1/4 3/12	10/46 5/23	1/2 15/30	
2/8 1/4	2/10 1/5	4/8 1/2	2/3 12/18	10/24 5/12
200/300 2/3	1/10 100/1000	3/8 6/16	6/9 2/3	

Solution: Adding Decimals

234.60 + 1.13 235.73	15.80 + 1.70 17.50	6.54 + 3.35 9.89	0.31 + 0.24 0.55	7.45 + 1.25 8.7
4.7 + 0.5 5.2	75.3 + 1.4 76.7	26.29 24.50 + 1.79	11.11 + 0.88 11.99	9.66 8.99 + 0.67	207.40 + 153.20 360.6
6.5 + 5.0 11.5	8.66 + 7.60 16.26	92.10 + 3.0 95.1	3.0 + 0.9 3.9	1.29 + 0.04 1.33
31.65 + 1.30 32.95	$3.25 $2.50 + $0.75	$0.42 + $0.03 forty-five cents	17.34 9.89 + 7.45	78.5 + 25.70 104.20	$1.10 + $0.40 $1.50
8.75 + 6.25 15	8.0 + 2.6 3.4	2.71 + 2.58 5.29	$15 + $5.25 20.25	4.75 + 4.05 8.8
0.60 + 0.20 0.80	0.98 + 0.02 one	$1.29 + $0.75 $2.04	99.80 + 78.19 177.99	7.9 4.9 + 3.0	1.2 + 1.1 2.3
$1.10 + $0.75 $1.85	$0.45 + $0.17 sixty-two cents	9.89 + 7.25 17.14	$68.24 + $68.20 $136.44	$4.58 + $4.55 $9.40
123.41 + 1.4 124.81	$2.21 + $1.25 $3.46	$93.65 86.45 + 7.20	4.64 + 0.21 4.85	5.25 + 4.9 10.15	7.12 + 0.58 7.7
$10 + $0.80 $10.80	1.12 + 1.05 2.17	$0.75 + $0.10 eighty-five cents	1.25 + 0.075 1.325	0.25 + 0.14 0.39

Solution: Subtracting Decimals

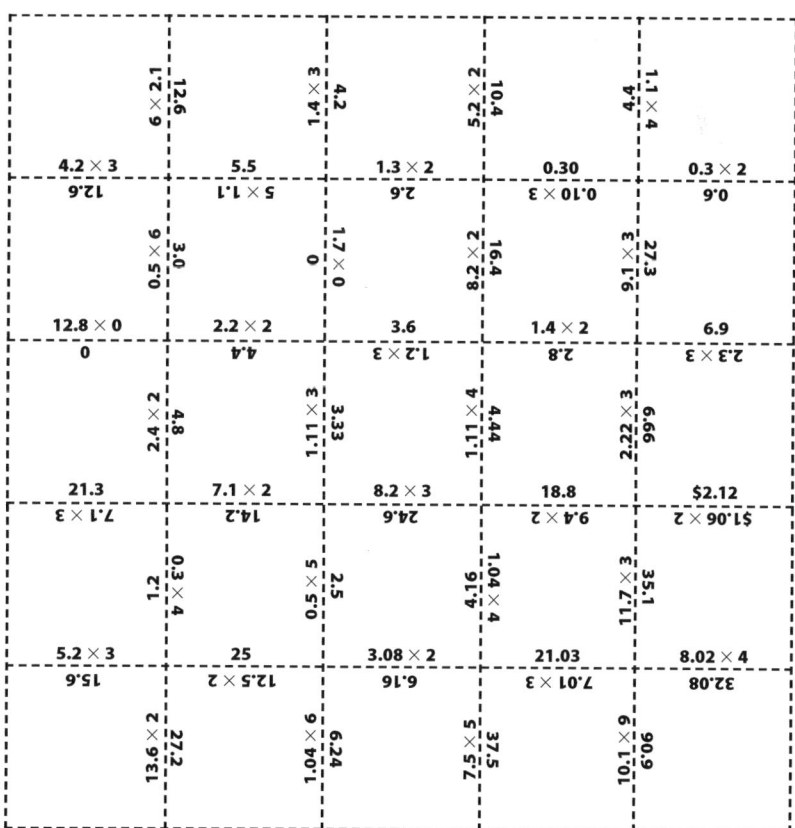

The puzzle grid contains the following problems and answers:

234.60 − 1.13	75.3 − 1.4	6.54 − 3.35	0.31 − 0.24	7.45 − 1.25	
4.2	**73.9**	**3.19**	**0.07**	**6.20**	**54.20**
4.7 − 0.5	6.5 − 5.0	92.10 − 3.0	3.0 − 0.9	1.29 − 0.04	207.40 − 153.20
31.65 − 1.30	**$2.50 − $0.75**	**$0.39**	**2.44**	**78.5 − 25.70**	**$1.10 − $0.40**
8.75 − 6.25	$2.71 − $1.79	0.42 − 0.03	9.89 − 7.45	$15 − $5.25	4.75 − 4.05
0.60 − 0.20	**$0.98 − $0.09**	**$1.29 − $0.75**	**21.61**	**4.9 − 3.0**	**0.1**
thirty-five cents	0.45 − 0.17	2.71 − 2.58	99.80 − 78.19	four dollars	$4.85 − $4.55	1.2 − 1.1
123.41 − 1.4	**$2.21 − $1.25**	**$86.45 − $7.20**	**1.43**	**5.25 − 4.9**	**7.12 − 0.58**
$10 − $0.80	1.12 − 1.05	sixty-five	1.25 − 0.75	0.25 − 0.14	

Solution: Multiplying Decimals

The puzzle grid contains the following problems and answers:

12.6	14.4 × 3	5.2 × 2	1.1 × 4	
4.2 × 3	**5.5**	**1.3 × 2**	**0.30**	**0.3 × 2**
9.6 × 0.5	1.7 × 0	8.2 × 2	3 × 1.9	
12.8 × 0	**2.2 × 2**	**3.6**	**1.4 × 2**	**6.9**
2.4 × 2	1.11 × 3	1.11 × 4	2.22 × 3	
21.3	**7.1 × 2**	**8.2 × 3**	**18.8**	**$2.12**
0.3 × 4	0.5 × 5	4.1 × 4	11.7 × 3	
5.2 × 3	**25**	**3.08 × 2**	**21.03**	**8.02 × 4**
13.6 × 2	1.04 × 6	7.5 × 5	10.1 × 9	

© 2005 Trish Caldwell Landsittel   **Puzzled by Math! 51**

Solution: Fraction and Decimal Conversion

eleven cents $0.11	1/5 0.20	sixty-eight hundredths 68/100	0.70 70/100	12 pennies $0.12
1/2 0.50	0.04 4/100	75 cents $0.75	3/4 0.75	0.7 7/10 · 2/4 1/2
one and six tenths 1.6	twenty cents 1/5 of $1.00	twelve cents 0.12	twenty-nine cents $0.29	9 dimes $0.90
0.25 1/4	3/10 0.3	2/2 1	0.33̄ 1/3	4/4 1 · nine and 2 tenths 9.2
nine hundredths 0.09	81/100 0.81	one dollar $1.00	80 cents $0.80	four hundredths 0.04
0.33 33/100	0.25 25/100	75/100 3/4	1.00 100/100	0.10 10/100 · 50/100 1/2
twenty-five cents $0.25	six tenths 0.6	1.5 3/2	2.0 4/2	one hundredth 0.01
0.75 75/100	0.77 77/100	0.80 80/100	6/10 0.6	1/10 of $1.00 10 cents · 1/2 5/10
2 dimes, 1 penny $0.21	5 nickels $0.25	2 quarters $0.50	2/6 1/3	fifty cents $0.50

greater than 0.4 ? 0.2	0.2 ? 2.0 less than	greater than 5.7 ? 3.6	greater than 8.2 ? 1.8	greater than 0.9 ? 0.6
$31.75 ? $31.25 greater than	4.1 ? 4.01 greater than	= 1 1/2 ? 1.5	4.4 ? 4.5 less than	0.7 ? 0.6 greater than · 7.1 ? 0.3 greater than
greater than 8.7 ? 8.6	10.1 ? 1.1 greater than	greater than 0.1 ? 0.6	5.7 ? 3.8 greater than	5.0 ? 5.9 less than
$0.50 ? $0.75 less than	2/10 ? 0.2 =	greater than $0.75 ? $0.25	1.25 ? 1.40 less than	5.3 ? 5.6 less than · = 7/100 ? 0.07
2.0 ? 0.2 greater than	1.3 ? 1.4 less than	greater than 5.3 ? 5.2	4.3 ? 4.5 less than	4.2 ? 3.9 greater than
23/100 ? 0.23 =	1/10 ? 0.10 =	greater than 0.34 ? 0.25	7.17 ? 7.56 less than	9.2 ? 6.4 greater than · = 3.5 ? 3.50
4.5 ? 3.7 greater than	0.5 ? 5.0 less than	4.8 ? 4.5 greater than	6.4 ? 2.8 greater than	3.9 ? 4.8 less than
0.60 ? 0.65 less than	greater than $2.50 ? $1.62	$7.63 ? $7.22 greater than	0.53 ? 0.47 greater than	greater than 1.3 ? 1.2 · 18.6 ? 18.2 greater than
3.6 ? 3.8 less than	11.2 ? 11.0 greater than	6.4 ? 6.9 less than	3.2 ? 0.9 greater than	8.2 ? 8.8 less than

Solution: Comparing Decimals (Greater Than, Less Than, and Equal To)

Solution: Basic Algebra

n + 8 = 16   n = 8

n + 19 = 30   n = 11

25x = 100   x = 4

350 - n = 200   n = 150

n + 15 = 25   4 = 15 - n   40n = 240   n = 1   n + 24 = 32

n = 10   n = 11   n = 6   129/n = 129   n = 8

7x = 56   x = 8   10 × n = 100   x = 25   4 × n = 20   n = 5

3x = 24   x = 8   60n = 1,200   4 × 100 = 400   n + 3 + 5 = 15

x = 8   2x = 16   n = 20   n = 100   n = 7

n = 7   8 × n = 56   120 - n = 80   3x = 36   5 + 6 + 7 + n = 20   n = 2

x = 5   3n = 21   4 + 8 + n = 45   7 + n = 15   n = 4

2x = 10   n = 7   n = 33   n = 8   4 + 9 + 3 + n = 20

36 - n = 6   n = 30   8x = 80   x = 10   x = 7   4x = 28   9n = 81   n = 9

30 - x = 10   n + 43 = 43   10 x = 1,000   n = 5   72 = n + 63

x = 20   n = 0   x = 100   7 + n = 12   n = 9

3n = 30   n = 10   12n = 120   n = 10   8n = 72   n = 9   5n = 50   n = 10

Solution: Algebra: Addition and Subtraction

n + 3 - 5 = 10   n = 12

20 - n + 3 = 10   n = 13

5 - 4 + x = 12   x = 11

50 - x + 40 + 40 = 80   x = 10

10 + 8 - n = 15   n = 3

n = 14   7 + 3 - n = 8   n = 10   7 + n - 3 = 10   n = 7   20 - 10 + n = 13

8 + n - 1 = 21   n = 2   n + 4 - 3 = 11   n = 6   15 - 8 + n = 14   n = 3

n + 20 - 3 = 25   n = 8   17 - 14 + x = 7   x = 4   3 + 4 - n = 6   n = 1   10 + 10 - n = 14   n = 14   450 + 50 - n = 250   n = 250

15 - n + 10 = 20   10 + 10 - n = 19   4 - x + 3 = 6   x = 0   x + 8 - 5 = 5   7 - x + 4 = 8

n = 5   n = 1   x = 1   3 + 7 + x = 10   x = 2   x = 3

21 = 15 + 15 - n   n = 9   n = 8   80 + 40 - 50 + n = 10   150 + 50 - n = 175   n = 25   300 + 200 - x = 100   x = 400

16 = x + 20 - 8   x = 100   x = 50   x - 500 + 0 = 500   n + 6 - 5 = 10   n = 12

x = 4   x - 50 + 25 = 75   100 - x + 0 = 50   x = 1,000   n = 9   n - 2 + 5 = 15

14 - x + 10 = 20   x = 4   18 - n + 6 = 18   n = 6   24 + 4 - x = 25   x = 3   30 + 20 - x = 45   x = 5   45 - 15 + n = 60   n = 30

n = 0   n = 8   n - 6 + 5 = 10   x = 9   21 = 14 + x - 0   x - 3 + 1 = 10

7 + 7 - n = 14   16 - n + 0 = 60   n = 11   20 - x + 4 = 15   x = 7   x = 12

2 + 8 + n = 5   n = 5   75 - 25 + n = 10   n = 10   40 - 30 + n = 15   n = 5   45 - 10 + n = 50   n = 15   15 - 5 + n = 21   n = 11

Solution: Squares

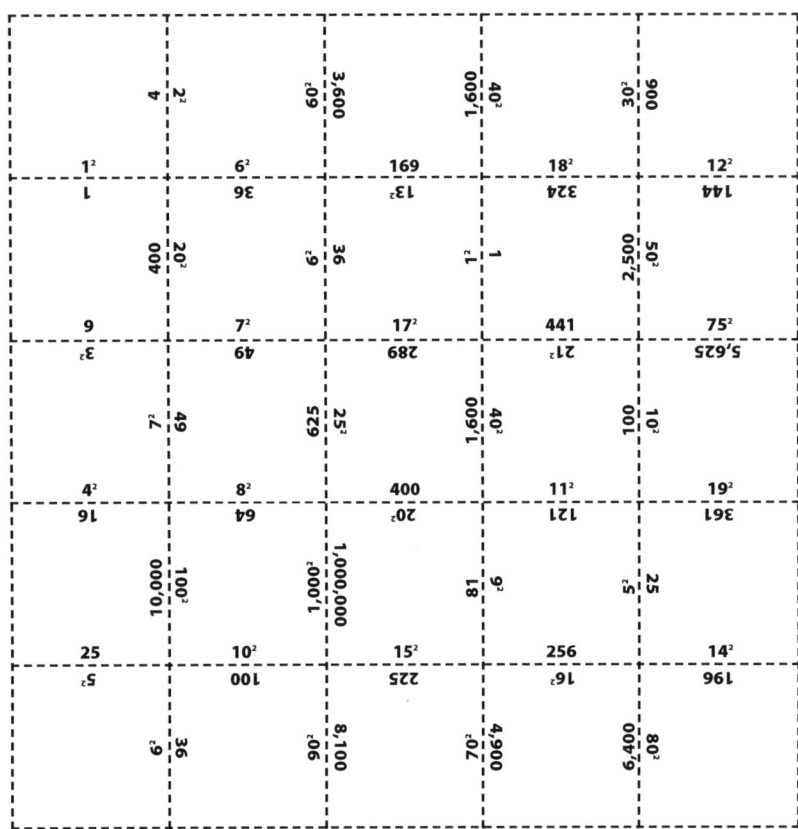

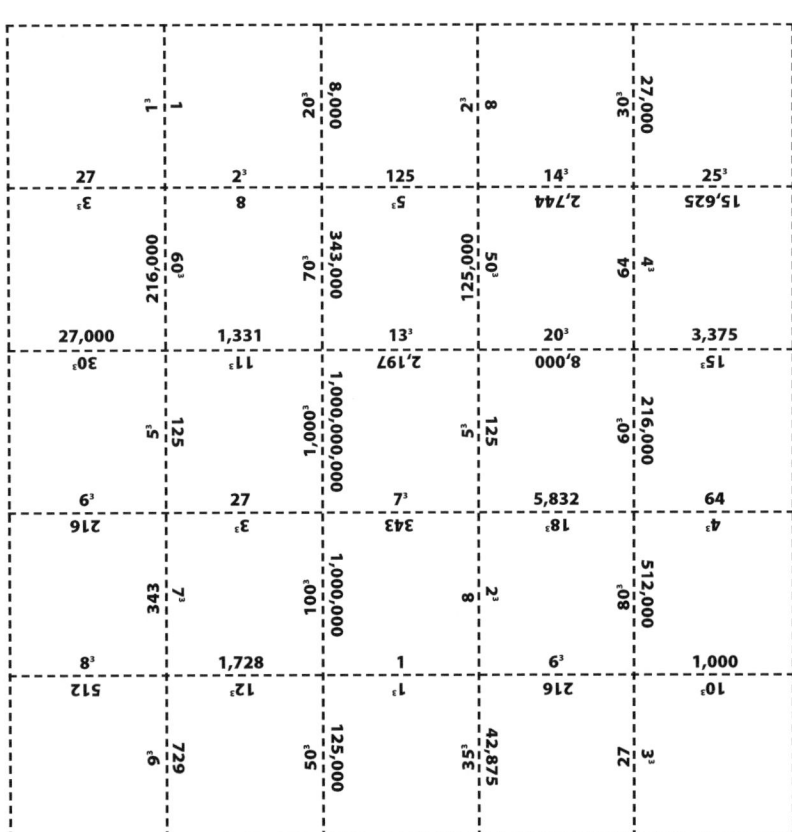

Solution: Cubes

Solution: Square Roots

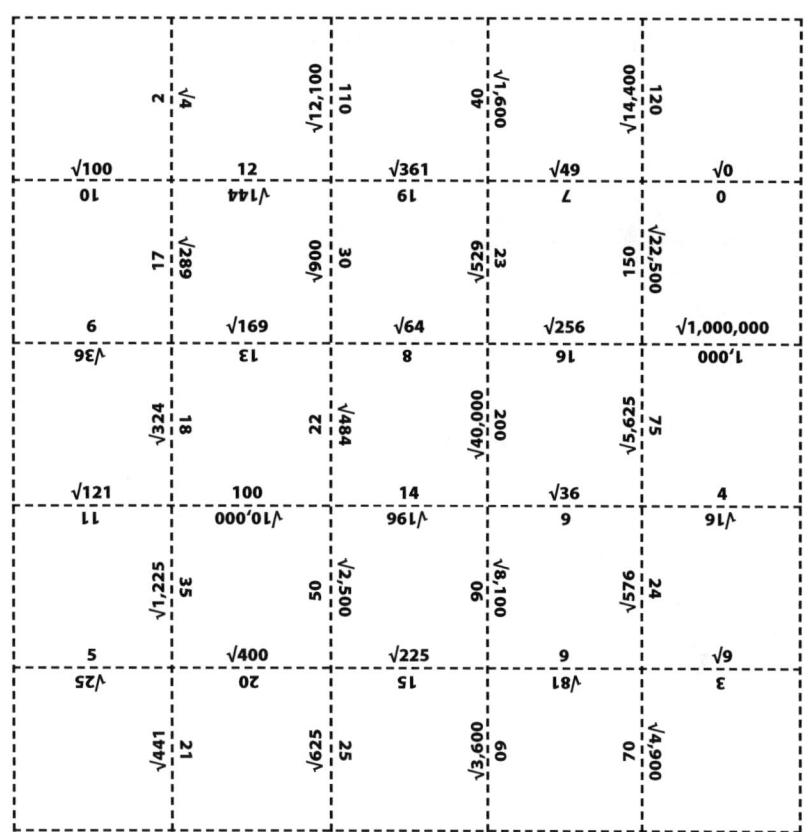

Appendix—Resources

24 Game
http://www.24game.com/

Continental Math League
PO Box 2196
St. James, NY 11780
http://www.continentalmathleague.hostrack.com/

Marcy Cook Math
Post Office Box 5840
Balboa Island, California 92662-5840
http://www.marcycookmath.com

Math Counts
1420 King Street
Alexandria, VA 22314
http://mathcounts.org/

Mathematics Pentathlon
1412 Sadlier Circle East Drive
Indianapolis, IN 46239
http://www.mathpentath.org

National Association for Gifted Children (NAGC)
1707 L Street NW Suite 550
Washington, DC 20036
http://www.nagc.org

National Council of Teachers of Mathematics (NCTM)
1906 Association Drive
Reston, VA 20191-1502
http://www.nctm.org

National Library of Virtual Manipulatives for Interactive Math
Utah State University
http://matti.usu.edu/nlvm/nav/index.html

The Math Forum @ Drexel (University)
3210 Cherry Street
Philadelphia, PA 19104
http://mathforum.org/

U.S. State Gifted Associations
http://www.gifted.uconn.edu/stategt.html

Blank Puzzle: 5 x 5

Blank Puzzle: 5 x 6

Blank Puzzle: 6 x 7

ACTIVITIES FOR ADVANCED LEARNING SERIES

Camp Fraction
Solving Exciting Word Problems Using Fractions
Set around a trip to summer camp, students work with fractions in a problem-solving format, while learning a little history, trivia, and fun facts about a number of different items.
Grades 4–6 $11.95

Creative Writing
Using Fairy Tales to Enrich Writing Skills
Use fairy tales to challenge and motivate your students. This activity book contains fun reading and writing activities that pique students' interest in creative writing.
Grades 4–8 $11.95

Extra! Extra!
Advanced Reading and Writing Activities for Language Arts
The book includes standards-based independent language arts activities for students in grades K–2 such as developing a newspaper and inventing new words.
Grades K-2 $11.95

Math Problem Solvers
Using Word Problems to Enhance Mathematical Problem Solving Skills
The standards-based problem solving strategies addressed in this book include drawing a picture, looking for a pattern, guessing and checking, acting it out, making a table or list, and working backwards.
Grades 2–3 $11.95

Puzzled by Math!
Using Puzzles to Teach Math Skills
Puzzled by Math! offers a collection of mathematical equations, knowledge, and skills in puzzle form. Standards-based content addresses addition, subtraction, multiplication, division, fractions, decimals, and algebra. Thirty-five exciting and challenging puzzles are included, as well as suggestions for using the material for a classroom learning center.
Grades 3–7 $11.95

Survival on the Reef
Exploring Amazing Animals and the Ways They Adapt to Their Environment
This challenging activity book addresses many essential skills and knowledge contained in the National Science Teachers Association standards using activities focused on the exciting environment of a coral reef, its inhabitants, and the ways these inhabitants have adapted to their world.
Grades 2–3 $11.95

Writing Stretchers
15 Minute Activities to Enrich Writing Skills
Standards-based activities address the areas of reading, writing, vocabulary, content literacy, creativity, and thinking skills, giving students a chance to enrich their writing skills.
Grades 4–8 $11.95

For a complete listing of titles in this series, please visit our website at

http://www.prufrock.com

PRUFROCK PRESS INC.